WebGIS系列丛书

WebGIS

之Element前端组件开发

郭明强 黄颖 杨亚仑 葛亮 高婷 王波 匡明星 曹威 宋振振 余晓敏 编著

電子工業出版社
Publishing House of Electronics Industry
北京•BEIJING

内 容 简 介

本书结合 Vue 和 OpenLayers，由浅入深、循序渐进地介绍 Element 的常用组件用法。本书共 8 章，首先介绍 Element+Vue+OpenLayers 开发环境的配置；然后结合 Vue 和 OpenLayers 对 Element 的常用组件进行详细的介绍，包括基本组件、表单组件、数据组件、通知组件、导航组件以及其他组件；最后以开发一个简单的智慧校园系统为例，进行 Element+Vue+OpenLayers 的项目实战。

本书既可作为高等学校计算机科学与技术、地理信息系统、网络 GIS、互联网软件开发、地理空间信息工程等相关专业或课程的教材和教学参考书，也可供计算机、GIS、遥感、测绘等领域的科研工作者参考。

本书提供了配套的开发程序代码，读者可登录华信教育资源网（www.hxedu.com.cn）免费注册后下载。

图书在版编目（CIP）数据

WebGIS 之 Element 前端组件开发 / 郭明强等编著. —北京：电子工业出版社，2021.12
（WebGIS 系列丛书）
ISBN 978-7-121-43162-3

Ⅰ. ①W… Ⅱ. ①郭… Ⅲ. ①地理信息系统－应用软件 Ⅳ. ①P208

中国版本图书馆 CIP 数据核字（2022）第 046188 号

责任编辑：田宏峰
印　　刷：天津千鹤文化传播有限公司
装　　订：天津千鹤文化传播有限公司
出版发行：电子工业出版社
　　　　　北京市海淀区万寿路 173 信箱　邮编 100036
开　　本：787×1 092　1/16　印张：16.5　字数：419 千字
版　　次：2021 年 12 月第 1 版
印　　次：2021 年 12 月第 1 次印刷
定　　价：99.00 元

凡所购买电子工业出版社图书有缺损问题，请向购买书店调换。若书店售缺，请与本社发行部联系，联系及邮购电话：（010）88254888，88258888。

质量投诉请发邮件至 zlts@phei.com.cn，盗版侵权举报请发邮件至 dbqq@phei.com.cn。

本书咨询联系方式：tianhf@phei.com.cn。

前　言

在互联网+、云计算、物联网、人工智能等新兴技术蓬勃发展的大背景下，智慧城市、智慧交通、智慧农业、智慧国土、智慧校园等各种互联网信息系统纷纷从桌面应用向互联网应用转变。基于 Web 的互联网应用具有易维护、易更新、易推广等特性，已成为广大互联网企业和程序开发者的首选方案。

目前流行的 Web 开发组件较多且各具特色，如 ExtJS、BootStrap、Element、jQuery UI 等。其中 Element 作为新兴的 Web 开发组件库，能较好地与 Vue 结合使用，提高 Web 应用开发的效率。本书结合 Vue 和 OpenLayers 介绍 Element 各种组件的使用方法。首先介绍 Element+Vue+OpenLayers 开发环境的配置，然后对 Element 的基本组件、表单组件、数据组件、通知组件、导航组件及其他组件进行介绍，最后以智慧校园系统为例进行 Element+Vue+OpenLayers 项目实战，为广大开发者提供了较为全面的学习资料。

本书主要面向计算机、GIS、遥感、测绘等相关领域的工作者，在内容编排上遵循一般学习规律，由浅入深、循序渐进地介绍 Element 组件的使用方法，内容完整、实用性强，既有详尽的代码阐述，又有丰富的图形展示，可以使读者快速、全面地掌握 Element 组件的使用方法。对于初学者来说，只需要按部就班地跟着本书进行学习即可。无论读者是否具有 Web 应用开发经验，都可以借助本书来系统地了解和掌握基于 Element 的 Web 应用开发所需的技术知识点，为开发新颖的 Web 应用奠定良好的基础。

本书提供了全部示例源码，其中第 2 章到第 8 章的每个示例都能独立运行，读者可快速查看演示效果。本书内容组织与 Element 官网的线上资源保持一致，读者在学习本书的过程中，可以查看 Element 官网提供的示例进行更深入的学习。

本书出版得到国家自然科学基金（41701446、41971356）的支持，在此表示诚挚的谢意。在本书的编写过程中，电子工业出版社的田宏峰编辑提出了宝贵的建议，在此表示感谢。

本书作者均长期从事网络地理信息系统的理论方法研究、教学和应用开发工作，有多年的网络地理信息系统和互联网软件开发经验。这为本书的编写打下了扎实的理论基础和实践基础。尽管如此，限于作者水平，本书难免会有不足和疏漏之处，敬请广大读者批评指正。

作　者

2021 年 7 月

前言

目　　录

第 1 章 开发环境的配置

1.1 使用 npm 配置开发环境

如果读者进行的开发是项目级别的，则需要使用 npm 来配置开发环境；如果读者是个初学者，则只需要使用直接引用的方式即可。本书第 1 章介绍开发环境的配置，第 2 章到第 7 章采用适合初学者的直接应用方式来介绍 Element 的常用组件，第 8 章使用 npm 配置的开发环境来实现一个项目级别的开发——智慧校园系统。

使用 npm 配置开发环境主要包括 Vue、Element 和 OpenLayers 的安装。

1.1.1 Vue 的安装

在安装 Vue 之前，需要先安装 node.js 和 vue-cli。

1．安装 node.js

从 node.js 官网下载 node，建议下载 LTS 版本（长期稳定版本），如图 1-1 所示。

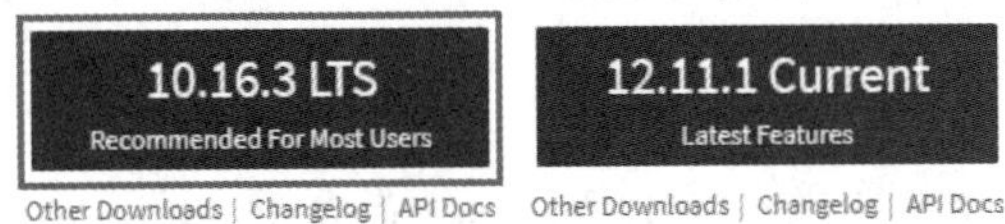

图 1-1　在 node.js 官网下载 node

Vue 的安装过程很简单，一直单击“下一步”按钮即可。安装完之后，在命令行窗口执行命令：

```
node -v
```

可查看 node 的版本，若在命令行窗口中显示相应的版本号，则说明 node 安装成功了，如图 1-2 所示。

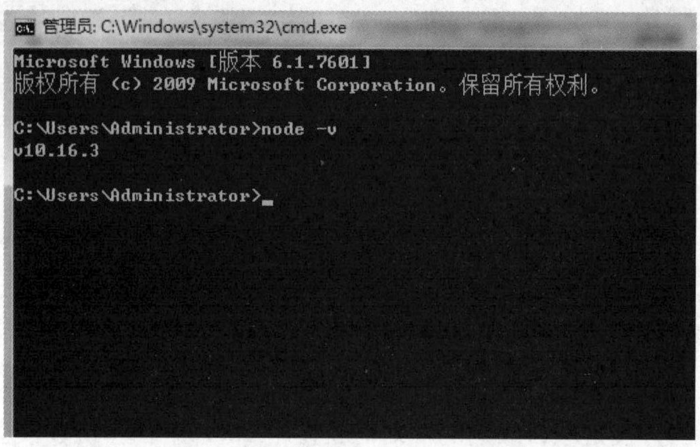

图 1-2　在命令行窗口显示 node 的版本号

npm 是 node 的包管理器，集成在 node 中，安装 node 后就有了 npm。直接在命令行窗口中执行命令：

```
npm -v
```

可看到 npm 的版本号，如图 1-3 所示。

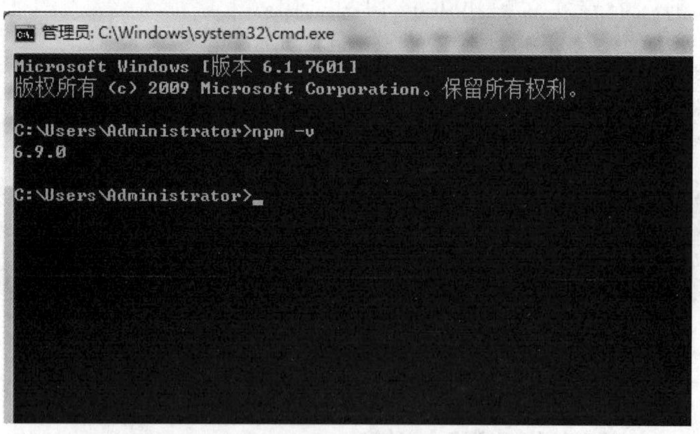

图 1-3　在命令行窗口显示 npm 的版本号

到目前为止，node 的开发环境已经安装完成，也有了 npm。由于有些 npm 资源被屏蔽了或者是在国外，所以会经常导致 npm 安装依赖包失败，因此还需要 npm 的国内镜像——cnpm。

在命令行窗口中执行命令：

```
npm install –g cnpm -- registry=http://registry.npm.taobao.org
```

如果没有报错，则表示 cnpm 安装成功。在命令行窗口中执行命令：

```
cnpm –v
```

可查看 cnpm 的版本号，如图 1-4 所示。

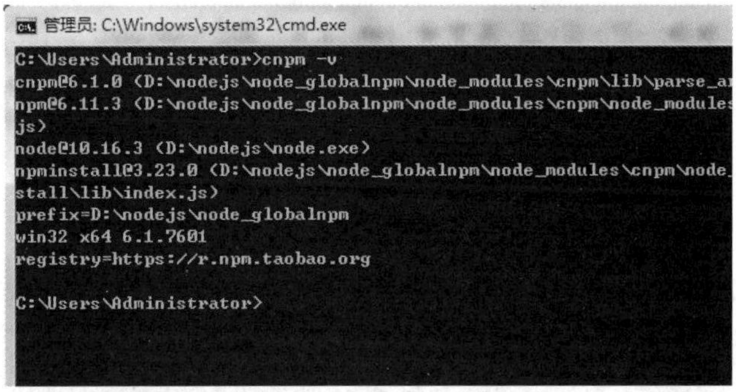

```
管理员: C:\Windows\system32\cmd.exe
C:\Users\Administrator>cnpm -v
cnpm@6.1.0 (D:\nodejs\node_globalnpm\node_modules\cnpm\lib\parse_a
npm@6.11.3 (D:\nodejs\node_globalnpm\node_modules\cnpm\node_module
js)
node@10.16.3 (D:\nodejs\node.exe)
npminstall@3.23.0 (D:\nodejs\node_globalnpm\node_modules\cnpm\node_
stall\lib\index.js)
prefix=D:\nodejs\node_globalnpm
win32 x64 6.1.7601
registry=https://r.npm.taobao.org

C:\Users\Administrator>
```

图 1-4　在命令行窗口显示 cnpm 的版本号

成功安装 cnpm 之后，就可以使用 cnpm 来安装依赖包了。如果想进一步了解 cnpm，可查看 TAONPM 官网。在下面的安装中，若安装速度过慢，则可将 npm 改为 cnpm。

2．安装 vue-cli 脚手架构建工具

在命令行窗口中执行命令：

```
npm install –g @vue/cli
```

或者

```
npm install –g vue-cli
```

前者安装的是 3.X 版本的 vue-cli，后者安装的是 2.X 版本的 vue-cli。在命令行窗口中执行命令：

```
vue –V
```

如果在命令行窗口中显示版本号，如图 1-5 所示，则表示 vue-cli 安装成功。注意，上面命令中的 V 要大写。

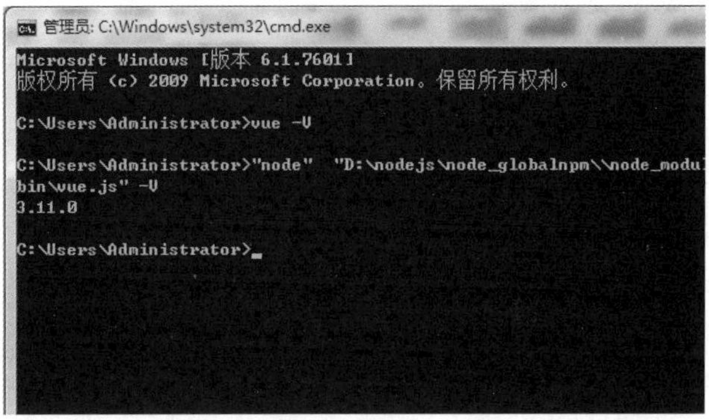

```
管理员: C:\Windows\system32\cmd.exe
Microsoft Windows [版本 6.1.7601]
版权所有 (c) 2009 Microsoft Corporation。保留所有权利。

C:\Users\Administrator>vue -V

C:\Users\Administrator>"node"  "D:\nodejs\node_globalnpm\\node_modu
bin\vue.js" -V
3.11.0

C:\Users\Administrator>
```

图 1-5　在命令行窗口显示 Vue 的版本号

2.X 版本的 vue-cli 只能使用命令行窗口来创建项目。首先在将要创建项目的地方按下 shift+鼠标右键，在弹出的右键菜单中选择“在此处打开命令行窗口”选项；然后在命令行窗口中执行命令：

```
vue init webpack firstApp,
```

firstApp 是要创建的文件夹名称；最后根据自己的需要在命令行窗口中设置项目。

3.X 版本的 vue-cli 可以使用可视化页面进行项目的创建，在命令行窗口执行命令：

```
vue ui
```

即可打开可视化页面，如图 1-6 所示。使用可视化页面创建项目如图 1-7 所示。

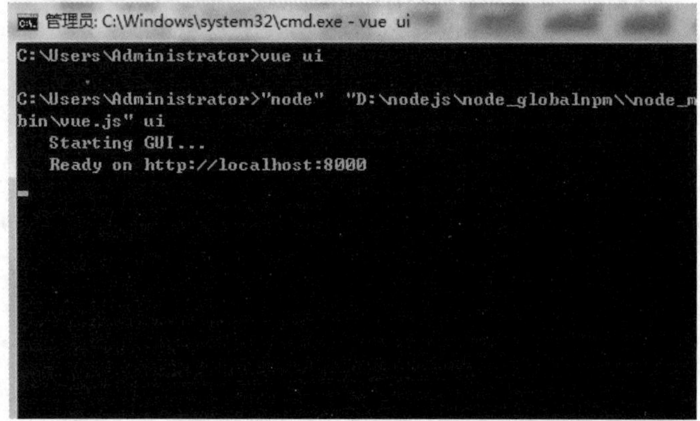

图 1-6　打开可视化页面的命令

图 1-7　使用可视化页面创建项目

3.X 版本的 vue-cli 也可以使用命令行窗口创建项目，但需要安装一个桥接工具，在命令行窗口中执行命令：

```
npm install –g@vue/cli-init
```

即可安装桥接工具。在桥接工具安装成功后，在命令行窗口中执行命令：

```
vue init webpack firstApp
```

即可创建项目。

创建成功的项目目录结构如图 1-8 所示。

图 1-8　创建成功的项目目录结构

图 1-8 中，文件夹 node_modules 中保存的是项目的依赖包文件，package.json 是项目依赖包的配置文件，该文件中的“dependencies”键（key）对应的值（value）是该项目引入的外部库文件，如图 1-9 所示。

```
{
  "name": "yyl",
  "version": "1.0.0",
  "description": "",
  "author": "杨亚仑 <531276717@qq.com>",
  "private": true,
  "scripts": {
    "dev": "webpack-dev-server --inline --progress --config build/web
    "start": "npm run dev",
    "unit": "jest --config test/unit/jest.conf.js --coverage",
    "test": "npm run unit",
    "build": "node build/build.js"
  },
  "dependencies": {
    "vue": "^2.5.2",
    "vue-router": "^3.0.1"
  },
  "devDependencies": {
    "autoprefixer": "^7.1.2",
    "babel-core": "^6.22.1",
    "babel-helper-vue-jsx-merge-props": "^2.0.3",
```

图 1-9　项目引入的外部库文件

项目成功创建后，可以在编辑器的终端或者项目根目录打开命令行窗口并执行命令：

```
npm run dev
```

或

```
npm run start
```

来运行项目。项目运行成功后，会有一个 IP 地址，浏览器中输入该 IP 即可打开创建的项目，打开项目时的初始化页面如图 1-10 所示。

图 1-10　打开项目时的初始化页面

需要注意的是，一定要全局安装 vue-cli 脚手架构建工具。全局安装是指在计算机的任何位置都可以访问 Vue，局部安装只能在局部位置访问 Vue，在命令行窗口中使用“-g”表示全局安装。如果采用全局安装，却不能在任何位置访问 Vue，那么将“D:\nodejs\node-globalnpm”（根据自己 node 的安装位置）添加到系统环境变量中就可以解决该问题。

1.1.2　Element 的安装

在项目的根目录中，首先按下 Shift+鼠标右键，在弹出的右键菜单中选择“在此处打开命令行窗口”；然后在命令行窗口中执行命令：

```
npm i element-ui -S
```

即可开始下载并安装 Element。成功安装 Element 后，打开 package.json 文件，可发现“dependencies”键所对应的值包含 element-ui，如图 1-11 所示。

```
    "unit": "jest --config test/unit/jest.conf.js --coverage",
    "test": "npm run unit",
    "build": "node build/build.js"
  },
  "dependencies": {
    "element-ui": "^2.12.0",
    "vue-router": "^3.0.1"
  },
  "devDependencies": {
    "autoprefixer": "^7.1.2",
    "babel-core": "^6.22.1",
    "babel-helper-vue-jsx-merge-props": "^2.0.3",
    "babel-jest": "^21.0.2",
    "babel-loader": "^7.1.1",
    "babel-plugin-dynamic-import-node": "^1.2.0",
    "babel-plugin-syntax-jsx": "^6.18.0",
    "babel-plugin-transform-es2015-modules-commonjs": "^6.26.0",
    "babel-plugin-transform-runtime": "^6.22.0",
```

图 1-11　“dependencies”键所对应的值包含 element-ui

1.1.3　OpenLayers 的安装

在项目的根目录中，按下 Shift+鼠标右键，在弹出的右键菜单中选择“在此处打开命令行窗口”，然后在命令行窗口中执行命令：

```
npm i ol -S
```

即可开始下载并安装 OpenLayers。成功安装 OpenLayers 后，打开 package.json 文件，可发现“dependencies”键所对应的值包含 ol，如图 1-12 所示。

```
    "unit": "jest --config test/unit/jest.conf.js --coverage",
    "test": "npm run unit",
    "build": "node build/build.js"
  },
  "dependencies": {
    "element-ui": "^2.12.0",
    "ol": "^6.0.0",
    "vue-router": "^3.0.1"
  },
  "devDependencies": {
    "autoprefixer": "^7.1.2",
    "babel-core": "^6.22.1",
    "babel-helper-vue-jsx-merge-props": "^2.0.3",
    "babel-jest": "^21.0.2",
    "babel-loader": "^7.1.1",
    "babel-plugin-dynamic-import-node": "^1.2.0",
    "babel-plugin-syntax-jsx": "^6.18.0",
    "babel-plugin-transform-es2015-modules-commonjs": "^6.26.0",
    "babel-plugin-transform-runtime": "^6.22.0",
    "babel-plugin-transform-vue-jsx": "^3.5.0",
    "babel-preset-env": "^1.3.2",
    "babel-preset-stage-2": "^6.22.0",
```

图 1-12　“dependencies”键所对应的值包含 ol

1.2　采用直接引用的方式配置开发环境

对于初学者来讲，采用直接引用的方式配置开发环境是最简单、最容易理解的，直接通过<script>标签在 HTML 页面中引用所需要的库即可。本节主要介绍直接引用 Vue、Element、OpenLayers 的方法。

1.2.1　下载 Vue 文件

打开 Vue 官网，首先单击“起步”按钮，然后单击页面中的“安装”按钮，接着找到页面中的“开发版本”按钮和“生产版本”按钮，最后单击“开发版本”按钮即可将 Vue 文件下载到本地，如图 1-13 所示。

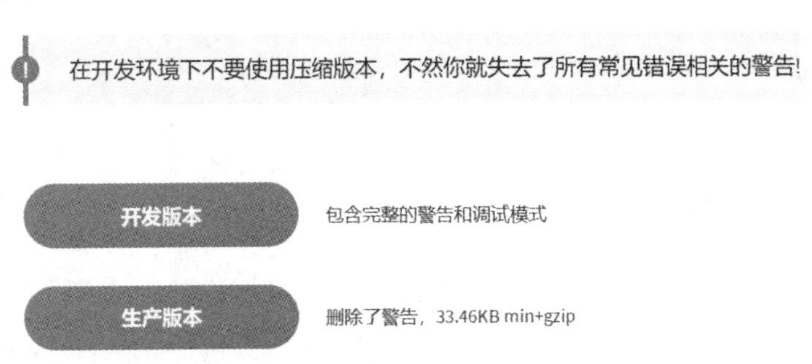

图 1-13　单击“开发版本”按钮下载 Vue 文件

读者也可以使用 CDN 的方式直接在线引用 Vue 文件，如图 1-14 所示。

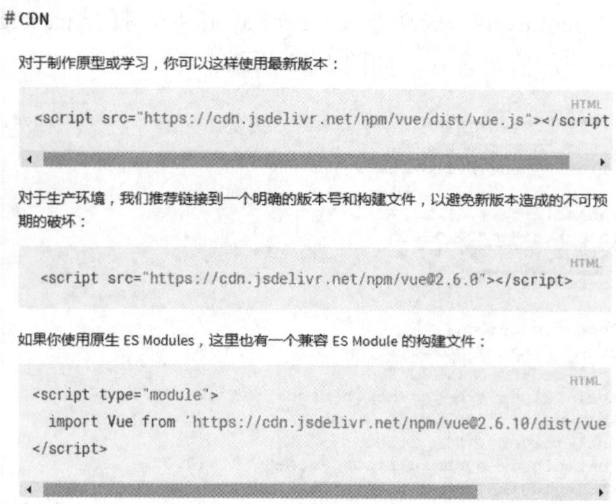

图 1-14　使用 CDN 的方式直接在线引用 Vue 文件

1.2.2　下载 Element 文件

打开 Element 官网，首先单击官网首页的“组件”按钮，然后在新页面中单击“unpkg.com/element-ui”，如图 1-15 所示，此时可弹出 UNPKG 页面，如图 1-16 所示。

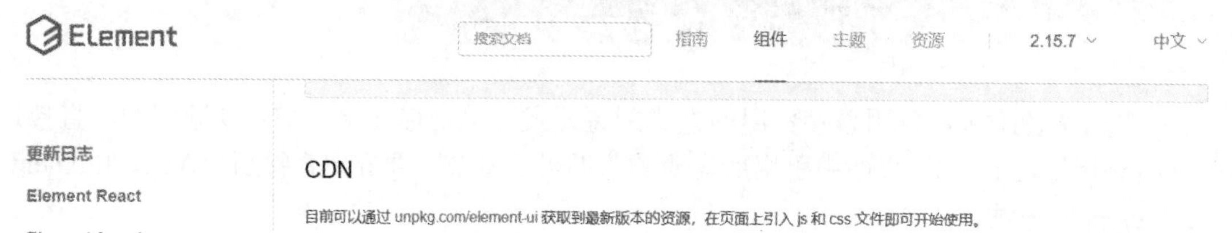

图 1-15　单击“组件”按钮后单击“unpkg.com/element-ui”

UNPKG

element-ui　　Version: 2.15.6

7 files, 4 folders		
lib	-	-
packages	-	-
src	-	-
types	-	-
CHANGELOG.en-US.md	69.6 kB	text/markdown
CHANGELOG.es.md	82.6 kB	text/markdown
CHANGELOG.fr-FR.md	82.7 kB	text/markdown
CHANGELOG.zh-CN.md	70.6 kB	text/markdown
LICENSE	1.08 kB	text/plain
README.md	6.34 kB	text/markdown
package.json	5.59 kB	application/json

图 1-16　UNPKG 页面

打开图 1-16 中的文件夹 lib 中的文件 index.js，如图 1-17 所示。单击图 1-17 中的“View Raw”按钮，可得到 Element 的 JavaScript 代码，如图 1-18 所示（只显示了部分代码）。

图 1-17　打开文件 index.js 后的页面

图 1-18　Element 的 JavaScript 代码

按下组合键 Ctrl+S，可将该页面的 JavaScript 代码保存到本地。打开“lib\theme-chalk”中的文件 index.css，单击页面中的“View Raw”按钮，按下组合键 Ctrl+S 将 index.css 的 JavaScript 代码保存到本地。

读者也可以在 Element 的官网中，通过 CDN 的方式直接引用 Element 文件，如图 1-19 所示。

图 1-19　通过 CDN 的方式直接引用 Element 文件

1.2.3 下载 OpenLayers 文件

打开 OpenLayers 官网，找到并单击“Get the Code”，如图 1-20 所示。在打开的页面中单击“v6.0.0-dist.zip”，如图 1-21 所示，即可下载 OpenLayers 文件的压缩包，解压缩后，将其中的 ol.css 文件和 ol.js 文件复制到自己的项目中。

Get the Code

Get the latest release or dig through the archives.

图 1-20 单击“Get the Code”

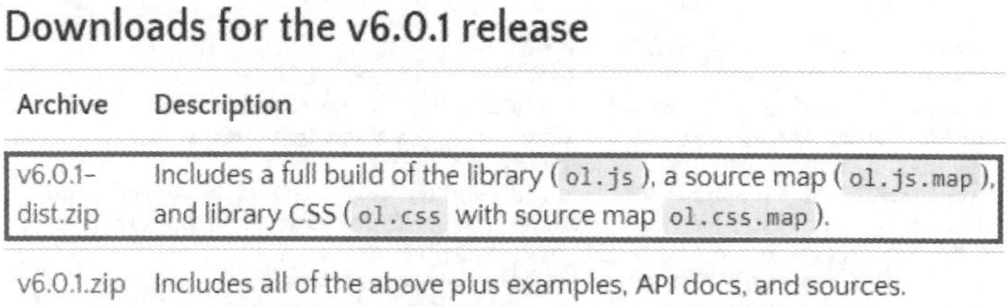

Downloads for the v6.0.1 release

Archive	Description
v6.0.1-dist.zip	Includes a full build of the library (ol.js), a source map (ol.js.map), and library CSS (ol.css with source map ol.css.map).
v6.0.1.zip	Includes all of the above plus examples, API docs, and sources.

图 1-21 单击“v6.0.0-dist.zip”

读者也可以直接引用 OpenLayers 文件，如图 1-22 所示。

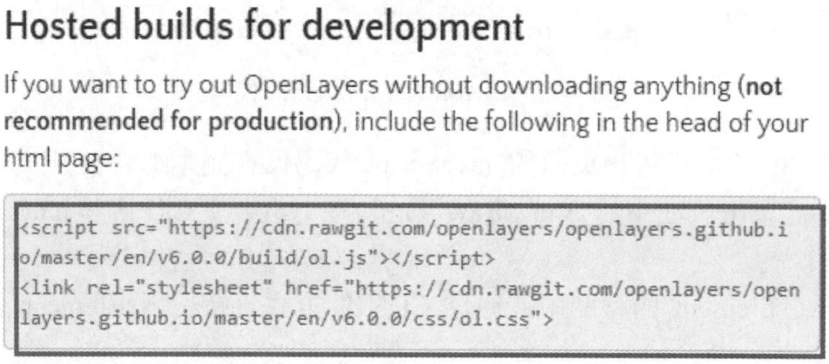

Hosted builds for development

If you want to try out OpenLayers without downloading anything (**not recommended for production**), include the following in the head of your html page:

```
<script src="https://cdn.rawgit.com/openlayers/openlayers.github.io/master/en/v6.0.0/build/ol.js"></script>
<link rel="stylesheet" href="https://cdn.rawgit.com/openlayers/openlayers.github.io/master/en/v6.0.0/css/ol.css">
```

图 1-22 直接引用 OpenLayers 文件

此时，我们可以创建一个名为 app 的文件夹，首先在该文件夹下创建一个文件夹 lib，用于存放项目需要引用的本地资源；然后在 lib 文件夹中创建 Vue 文件夹、Element 文件夹、OpenLayers 文件夹，将下载的 Vue 文件、Element 文件和 OpenLayers 文件分别存放在对应的文件夹中。项目的目录结构如图 1-23 所示，读者也可以按照个人习惯自定义项目的目录结构。

名称	修改日期	类型
Element	2019/9/30 17:13	文件夹
OpenLayers	2019/10/1 22:24	文件夹
Vue	2019/9/30 16:07	文件夹

图 1-23 项目的目录结构

选择一款适合自己的编辑器（本书使用的编辑器是 Visual Studio Code），在 app 文件夹中创建一个 HTML 文件，将刚刚下载的各种库文件引入到 HTML 文件中，如图 1-24 所示。

```
<meta charset="UTF-8">
<meta name="viewport" content="width=device-width, initial-scale=1.0">
<meta http-equiv="X-UA-Compatible" content="ie=edge">
<title>Document</title>
<link rel="stylesheet" href="lib\OpenLayers\ol.css">
<link rel="stylesheet" href="lib\Element\index.css">
<script src="lib\Vue\vue.js"></script>
<script src="lib\Element\index.js"></script>
<script src="lib\OpenLayers\ol.js"></script>
</head>
```

图 1-24　在 HTML 文件中引入各种库文件

1.3　第一个 Element+Vue+OpenLayers 示例

本节以某校园的遥感影像作为底图，通过 OpenLayers 将遥感影像加载到浏览器中，如图 1-25 所示。

单击“Button”按钮可弹出一个对话框，该对话框的内容为“Hello world”，如图 1-26 所示。

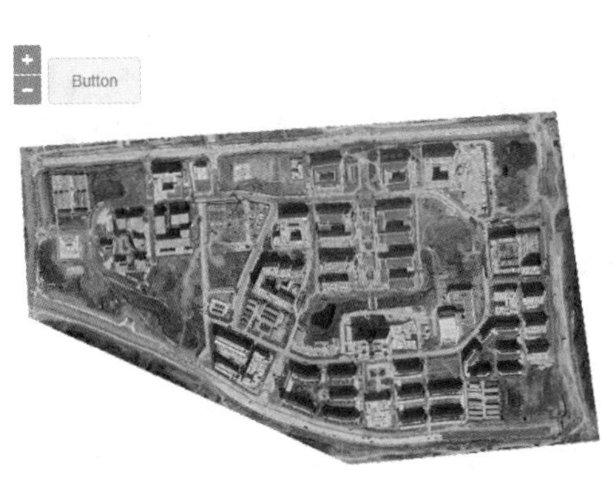

图 1-25　通过 OpenLayers 将遥感影像加载到浏览器中

图 1-26　内容为“Hello world”的对话框

图 1-26 所示对话框的实现代码如下：

```
<div id="app">
    <div id="map" class="map"></div>
    <!-- 通过 Element 的 Button 和 Dialog 标签定义出一个按钮和一个对话框 -->
    <el-button @click="visible = true" class="btn">Button</el-button>
    <el-dialog :visible.sync="visible" title="Hello world"></el-dialog>
</div>
```

```
<script>
    //定义一个 Vue 实例
    var vm = new Vue({
        el: "#app",
        data: function () {
            return { visible: false }
        },
        mounted: function () {
            //定义一个范围，该范围为图片的大小
            var extent = [0, 0, 4903, 2895];
            //定义一个投影
            var projection = new ol.proj.Projection({
                code: 'EPSG:4326',
                extent: extent
            })
            //通过 ol.layer.Image 定义一个图层，url 为该图片的地址
            var imageLayer = new ol.layer.Image({
                source: new ol.source.ImageStatic({
                    url: '../img/school.jpg',
                    projection: projection,
                    imageExtent: extent
                })
            })
            //定义一个 map，将 target、layers、view 所对应的值分别添加进去
            var map = new ol.Map({
                target: 'map',
                layers: [imageLayer],
                view: new ol.View({
                    projection: projection,
                    center: ol.extent.getCenter(extent),
                    zoom: 3
                })
            })
        }
    })
</script>
```

1.4 多语言切换

Element 组件内部默认使用的是中文，若希望使用其他语言，则需要进行多语言设置，通过 CDN 的方式可以在线加载语言文件。例如，若要使用英文，则需要引用<script src="//unpkg.com/element-ui/lib/umd/locale/en.js"></script>；若要使用中文，则需要引入<script src="//unpkg.com/element-ui/lib/umd/locale/zh-CN.js"></script>。读者可以在<script>中通过代码“ELEMENT.locale(ELEMENT.lang.zh-CN)或者 ELEMENT.locale(ELEMENT.lang.en)”来选择

使用中文或英文。Element 共内置了简体中文（zh-CN）、英语（en）、德语（de）等 60 种语言，读者可根据自己的需求来配置所需的语言。

当然，除了在线的 CDN 语言文件加载方式，读者还可以将语言下载到本地进行引用。在 Element 官网的首页中单击“组件”按钮，然后单击“unpkg.com/element-ui”，在弹出的页面中打开文件夹“lib\umd\locale”，可以看到多种语言的配置文件，如图 1-27 所示（该图只显示了部分语言的配置文件）。

UNPKG

element-ui / lib / umd / **locale**　　Version: 2.12.0

52 files		
..		
af-ZA.js	3.44 kB	application/javascript
ar.js	3.9 kB	application/javascript
bg.js	3.97 kB	application/javascript
ca.js	3.46 kB	application/javascript
cs-CZ.js	3.48 kB	application/javascript
da.js	3.33 kB	application/javascript

图 1-27　各种语言的配置文件

打开文件 zh-CN.js，可弹出文件 zh-CN.js 的页面，如图 1-28 所示。单击“View Raw”按钮，可看到文件 zh-CN.js 的 JavaScript 代码，如图 1-29 所示。按下组合键 Ctrl+S，可将文件 zh-CN.js 的 JavaScript 代码保存到本地，并将其添加到项目中，在 HTML 页面中引用文件 zh-CN.js 即可。

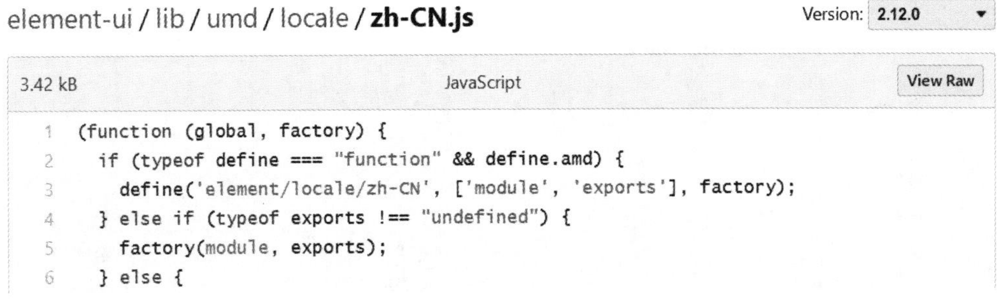

图 1-28　文件 zh-CN.js 的页面

```
(function (global, factory) {
  if (typeof define === "function" && define.amd) {
    define('element/locale/zh-CN', ['module', 'exports'], factory);
  } else if (typeof exports !== "undefined") {
    factory(module, exports);
  } else {
    var mod = {
      exports: {}
    };
    factory(mod, mod.exports);
    global.ELEMENT.lang = global.ELEMENT.lang || {};
    global.ELEMENT.lang.zhCN = mod.exports;
  }
})(this, function (module, exports) {
  'use strict';

  exports.__esModule = true;
  exports.default = {
    el: {
      colorpicker: {
        confirm: '确定',
        clear: '清空'
      },
      datepicker: {
        now: '此刻',
        today: '今天',
        cancel: '取消',
        clear: '清空',
        confirm: '确定',
        selectDate: '选择日期',
        selectTime: '选择时间',
        startDate: '开始日期',
```

图 1-29 文件 zh-CN.js 的 JavaScript 代码

1.5 自定义主题样式切换

Element 提供了一套默认使用的主题，Element 也提供使用其他自定义主题的方法，详见 Element 官网首页中的“主题”部分。本节通过直接引用的方法来自定义主题。

在 Element 官网的首页中单击“组件”按钮，然后单击“unpkg.com/element-ui”，在弹出的页面中打开文件夹“lib\theme-chalk”，如图 1-30 所示。

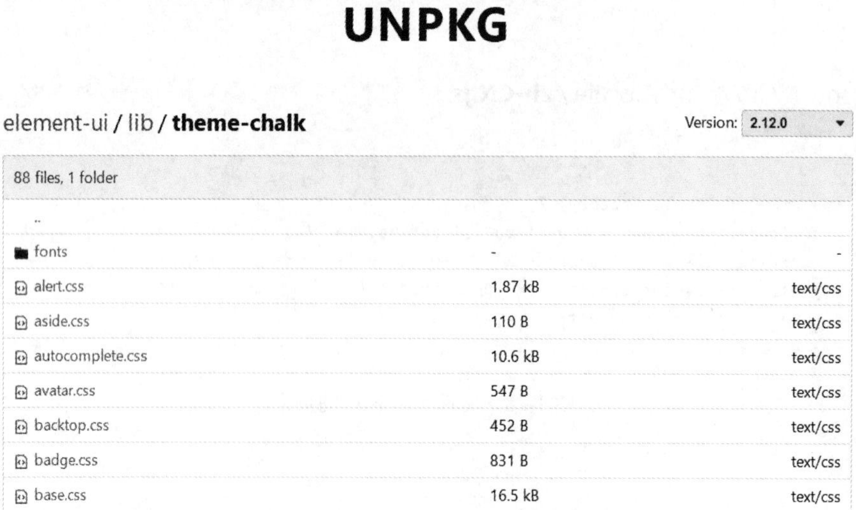

UNPKG

element-ui / lib / **theme-chalk** Version: 2.12.0

88 files, 1 folder		
..		
fonts	-	-
alert.css	1.87 kB	text/css
aside.css	110 B	text/css
autocomplete.css	10.6 kB	text/css
avatar.css	547 B	text/css
backtop.css	452 B	text/css
badge.css	831 B	text/css
base.css	16.5 kB	text/css

图 1-30 文件夹 theme-chalk 的页面

单击文件夹 theme-chalk 中的某个文件，如文件 autocomplete.css，可打开该文件的页面，如图 1-31 所示。单击图 1-31 中的“View Raw”按钮，可看到文件 autocomplete.css 的 JavaScript 代码，如图 1-32 所示（只显示了部分 JavaScript 代码）。按下组合键 Ctrl+S，可将文件 autocomplete.css 的 JavaScript 代码保存到本地，并将其添加到项目中，通过<link></link>标签将文件 autocomplete.css 引入 HTML 页面即可使用。

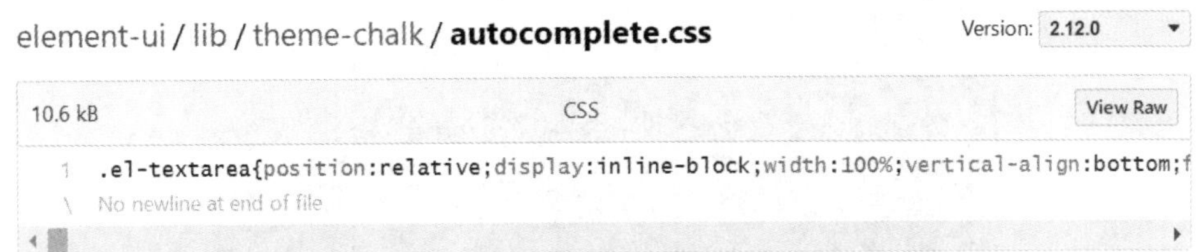

图 1-31　文件 autocomplete.css 的页面

```
.el-textarea{position:relative;display:inline-block;width:100%;vertical-
align:bottom;font-size:14px}.el-
textarea__inner{display:block;resize:vertical;padding:5px 15px;line-height:1.5;-
webkit-box-sizing:border-box;box-sizing:border-box;width:100%;font-
size:inherit;color:#606266;background-color:#FFF;background-
image:none;border:1px solid #DCDFE6;border-radius:4px;-webkit-transition:border-
color .2s cubic-bezier(.645,.045,.355,1);transition:border-color .2s cubic-
bezier(.645,.045,.355,1)}.el-textarea__inner::-webkit-input-
placeholder{color:#C0C4CC}.el-textarea__inner:-ms-input-
placeholder{color:#C0C4CC}.el-textarea__inner::-ms-input-
placeholder{color:#C0C4CC}.el-textarea__inner::placeholder{color:#C0C4CC}.el-
textarea__inner:hover{border-color:#C0C4CC}.el-
textarea__inner:focus{outline:0;border-color:#409EFF}.el-textarea .el-
input__count{color:#909399;background:#FFF;position:absolute;font-
size:12px;bottom:5px;right:10px}.el-textarea.is-disabled .el-
```

图 1-32　文件 autocomplete.css 的 JavaScript 代码

1.6　组件过渡动画

Element 提供的动画过渡效果有淡入淡出、缩放、展开折叠等。其中，淡入淡出和缩放使用的组件是<transition></transition>，而淡入淡出有 el-fade-in-linear 和 el-fade-in 两种效果，缩放有 el-zoom-in-center、el-zoom-in-top 和 el-zoom-in-bottom 三种效果。读者将 transition 组件中 name 的属性修改成不同的名字（如 el-fade-in-linear）即可实现相应的效果。展开折叠使用的组件为<el-collapse-transition>< /el-collapse-transition>。组件的动画过渡效果实例如图 1-33 所示。

当单击图 1-33 中的“Toggle”按钮，各个标签将以不同的动画效果隐藏起来（见图 1-34）；再次单击“Toggle”按钮，这些标签又会以不同的动画效果显示出来。

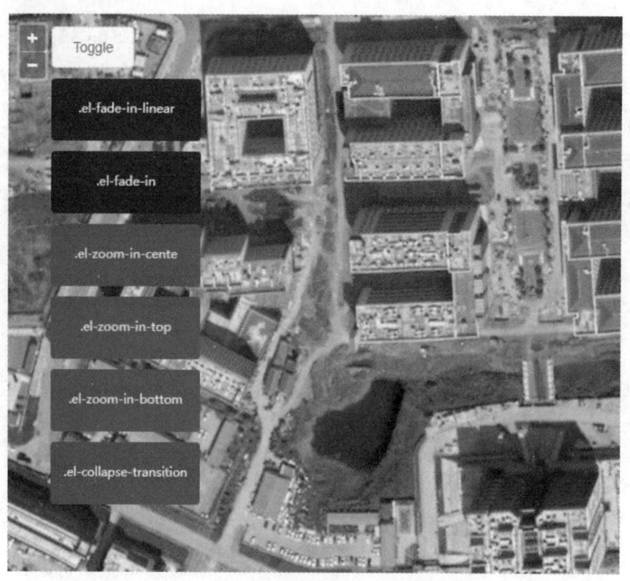

图 1-33　组件的动画过渡效果实例

图 1-34　以不同的动画效果隐藏各个标签

上述动画效果的实现代码如下：

```
<div class="container">
    <!-- Element 中的按钮，用于切换下面两个 div 的显示与隐藏 -->
    <el-button @click="toggleHandleClick" style="margin-bottom: 10px">Toggle</el-button>
    <!-- 淡入淡出动画（组件：transition，name：el-fade-in-linear） -->
    <transition name="el-fade-in-linear">
        <div v-show="show" class="transition-box color1" >.el-fade-in-linear</div>
    </transition>
    <!-- 淡入淡出动画（组件：transition，name：el-fade-in） -->
    <transition name="el-fade-in">
```

```
        <div v-show="show" class="transition-box color1">.el-fade-in</div>
    </transition>
    <!-- 缩放动画（组件：transition，name：el-zoom-in-cente） -->
    <transition name="el-zoom-in-center">
        <div v-show="show" class="transition-box color2">.el-zoom-in-cente</div>
    </transition>
    <!-- 缩放动画（组件：transition，name：el-zoom-in-top） -->
    <transition name="el-zoom-in-top">
        <div v-show="show" class="transition-box color2">.el-zoom-in-top</div>
    </transition>
    <!-- 缩放动画（组件：transition，name：el-zoom-in-bottom） -->
    <transition name="el-zoom-in-bottom">
        <div v-show="show" class="transition-box color2">.el-zoom-in-bottom</div>
    </transition>
    <!-- 展开折叠动画（组件:el-collapse-transition） -->
    <el-collapse-transition>
        <div v-show="show" class="transition-box color3">.el-collapse-transition</div>
    </el-collapse-transition>
</div>
```

第2章 基本组件

2.1 Element的布局

在搭建一个网页之前，首先要对网页的页面进行布局。CSS 原生布局在网页布局方面一直做得不是很好。CSS 原生布局使用的是 Table（表格）布局、Float（浮动）布局、Position（定位）布局和 Inline-block（行内块）布局等，但是这些布局方法在本质上还是 CSS Hack。CSS 原生布局后来添加了 Flexbox（盒子）布局，虽然解决了很多问题，但 Flexbox 布局是一维布局，不是二维布局。Grid（网格）布局的出现解决了复杂二维布局的问题，而且 Flexbox 布局可以和 Grid 布局混合使用，二者可以很好地配合。Element 封装了一套自己的页面布局方法，也可以和 Flexbox 配合，完成复杂的页面布局。

本节主要介绍 Element 的基础布局、分栏间隔、混合布局、分栏偏移、对齐方式等内容。

2.1.1 基础布局

使用 Element 搭建一个最基础的页面布局（基础布局），如图 2-1 所示。

图 2-1 基础布局

在基础布局中，页面被分为 24 栏，可在 el-col 标签中调整 span 属性来控制页面的列数。代码如下：

```
<div id="app">
    <div id="map" class="map"></div>
    <div class="container">
        <!-- 第一行里面有一列 -->
        <el-row>
            <el-col :span="24">
                <div class="grid-content color1"></div>
            </el-col>
        </el-row>
        <!-- 第二行里面有两列 -->
        <el-row>
            <el-col :span="12">
                <div class="grid-content color1"></div>
            </el-col>
            <el-col :span="12">
                <div class="grid-content color2"></div>
            </el-col>
        </el-row>
        <!-- 第三行里面有三列 -->
        <el-row>
            <el-col :span="8">
                <div class="grid-content color1"></div>
            </el-col>
            <el-col :span="8">
                <div class="grid-content color2"></div>
            </el-col>
            <el-col :span="8">
                <div class="grid-content color3"></div>
            </el-col>
        </el-row>
    </div>
</div>
```

2.1.2 分栏间隔

在基础布局的基础上可进行分栏间隔，分栏间隔是指各个盒子之间有一定的间距，如图 2-2 所示。

在使用 Element 进行分栏间隔时，只需要在 el-row 标签中的 gutter 属性中填写相应的间隔距离即可。代码如下：

```
<div id="app" class="app">
    <div id="map" class="map"></div>
    <div class="container">
        <!-- 定义一个一行四列的布局，通过 gutter 属性来控制每一列之间的距离为 20px -->
```

```
        <el-row :gutter="20">
            <el-col :span="6">
                <div class="grid-content color1"></div>
            </el-col>
            <el-col :span="6">
                <div class="grid-content color2"></div>
            </el-col>
            <el-col :span="6">
                <div class="grid-content color3"></div>
            </el-col>
            <el-col :span="6">
                <div class="grid-content color4"></div>
            </el-col>
        </el-row>
    </div>
</div>
```

图 2-2　分栏间隔

2.1.3　混合布局

混合布局是在基础布局和分栏间隔基础上实现的一种更为复杂的布局，如图 2-3 所示。

图 2-3　混合布局

有了基础布局和分栏间隔的知识基础，在使用 Element 进行页面布局时，可以将页面分为了 24 栏，其中 el-col 标签中的 span 属性可以控制该列所占的栏数，el-row 中的 gutter 属性表示每一列之间的间隔，单位为 px。代码如下：

```
<div id="app">
    <div id="map" class="map"></div>
    <div class="container">
        <!-- 第一行一共两列，第一列跨占 16 栏，第二列跨占 8 栏，每列间距 20px -->
        <el-row :gutter="20">
            <el-col :span="16">
                <div class="grid-content color1"></div>
            </el-col>
            <el-col :span="8">
                <div class="grid-content color1"></div>
            </el-col>
        </el-row>
        <!-- 第二行一共四列，第一列跨占 8 栏，第二列跨占 8 栏，第三列跨占 4 栏，第四列跨占 4
栏，每列间距 20px  -->
        <el-row :gutter="20">
            <el-col :span="8">
                <div class="grid-content color2"></div>
            </el-col>
            <el-col :span="8">
                <div class="grid-content color2"></div>
            </el-col>
            <el-col :span="4">
                <div class="grid-content color2"></div>
            </el-col>
            <el-col :span="4">
                <div class="grid-content color2"></div>
            </el-col>
        </el-row>
        <!-- 第三行一共三列，第一列跨占 4 栏，第二列跨占 16 栏，第三列跨占 4 栏，每列间距 20px  -->
        <el-row :gutter="20">
            <el-col :span="4">
                <div class="grid-content color3"></div>
            </el-col>
            <el-col :span="16">
                <div class="grid-content color3"></div>
            </el-col>
            <el-col :span="4">
                <div class="grid-content color3"></div>
            </el-col>
        </el-row>
    </div>
</div>
```

2.1.4　分栏偏移

分栏偏移是指各列的间距以栏为单位进行布局，如图 2-4 所示。

图 2-4　分栏偏移

分栏偏移使用了 el-col 标签中的 offset 属性，代码如下：

```
<div id="app">
    <div id="map" class="map"></div>
    <div class="container">
        <!-- 第一行一共两列，每列跨占 6 栏，第一列贴近左侧，第二列通过 offset 属性向左偏移 6 栏 -->
        <el-row :gutter="20">
            <el-col :span="6">
                <div class="grid-content color1"></div>
            </el-col>
            <el-col :span="6" :offset="6">
                <div class="grid-content color1"></div>
            </el-col>
        </el-row>
        <!-- 第二行一共两列，每列跨占 6 栏，两列分别向左偏移 6 栏 -->
        <el-row :gutter="20">
            <el-col :span="6" :offset="6">
                <div class="grid-content color2"></div>
            </el-col>
            <el-col :span="6" :offset="6">
                <div class="grid-content color2"></div>
            </el-col>
        </el-row>
        <!-- 第三行含有一列，跨占 12 栏，向左偏移 6 栏 -->
        <el-row :gutter="20">
            <el-col :span="12" :offset="6">
                <div class="grid-content color3"></div>
            </el-col>
        </el-row>
```

```
        </div>
    </div>
```

2.1.5 对齐方式

Element 的对齐方式包括 start、center、end、space-between、space-around 等。其中，start 表示左对齐；center 表示居中；end 表示右对齐；space-between 表示两边对齐，在每个网格子项中间放置均等的空间，在始末两端没有空间；space-around 表示在每个网格子项中间放置均等的空间，在始末两端只有中间空间的一半。对齐方式的效果如图 2-5 所示，图中从上到下分别是 start、center、end、space-between、space-around 对齐方式的效果。

图 2-5 对齐方式

将 el-row 标签中的 type 属性修改为 flex 类型，便可通过 justify 属性调整对齐方式，代码如下：

```
<div id="app">
    <div id="map" class="map"></div>
    <div class="container">
<!-- 第一行将 el-row 的类型设置为 flex，默认是 start 对齐方式 -->
<el-row type="flex" >
    <el-col :span="6">
        <div class="grid-content color1"></div>
    </el-col>
    <el-col :span="6">
        <div class="grid-content color1"></div>
    </el-col>
    <el-col :span="6">
        <div class="grid-content color1"></div>
    </el-col>
</el-row>
<!-- 第二行将 el-row 的类型设置为 flex，通过 justify 属性将其修改为 center 对齐方式 -->
<el-row type="flex"    justify="center">
    <el-col :span="6">
        <div class="grid-content color2"></div>
```

```
        </el-col>
        <el-col :span="6">
            <div class="grid-content color2"></div>
        </el-col>
        <el-col :span="6">
            <div class="grid-content color2"></div>
        </el-col>
    </el-row>
    <!-- 第三行将 el-row 的类型设置为 flex，通过 justify 属性将其修改为 end 对齐方式 -->
    <el-row type="flex" justify="end">
        <el-col :span="6">
            <div class="grid-content color3"></div>
        </el-col>
        <el-col :span="6">
            <div class="grid-content color3"></div>
        </el-col>
        <el-col :span="6">
            <div class="grid-content color3"></div>
        </el-col>
    </el-row>
    <!-- 第四行将 el-row 的类型设置为 flex，通过 justify 属性将其修改为 space-between 对齐方式 -->
    <el-row type="flex" justify="space-between">
        <el-col :span="6">
            <div class="grid-content color4"></div>
        </el-col>
        <el-col :span="6">
            <div class="grid-content color4"></div>
        </el-col>
        <el-col :span="6">
            <div class="grid-content color4"></div>
        </el-col>
    </el-row>
    <!-- 第五行将 el-row 的类型设置为 flex，通过 justify 属性将其修改为 space-around 对齐方式 -->
    <el-row type="flex"  justify="space-around">
        <el-col :span="6">
            <div class="grid-content color5"></div>
        </el-col>
        <el-col :span="6">
            <div class="grid-content color5"></div>
        </el-col>
        <el-col :span="6">
            <div class="grid-content color5"></div>
        </el-col>
    </el-row>
        </div>
    </div>
```

2.2 Element 的布局容器

在进行前端开发时，首先要考虑网页的整体布局。就像一位成熟的画家，首先要从大局入手，绘制出一个轮廓，再对各个局部进行修饰。

我们可以把一个网页想象成一个容器，为了能够把自己的物件更加井井有序地摆放在容器里，就必须对容器的空间进行划分。好的布局不仅能够使网页看起来更加美观，还能够提高开发效率。在 WebGIS 开发中，网页的布局与纯前端相比，没有那么复杂，但也不能忽略它的重要性。

在网页整体布局方面，Element 常用的标签是 el-container、el-header、el-aside、el-main 和 el-footer。其中，el-container 是一个容器级别的标签，其他标签将作为“槽”对该容器进行划分。需要注意的是 el-container 的子元素只能是后四者，后四者的父元素也只能是 el-container，并且当子元素包含 el-header 或 el-footer 时，全部子元素会垂直上下排列，否则子元素将水平左右排列。

本节将基于 Element 的布局样式和一张地图给出四种布局方式，对页面的整体布局进行详细的阐述。

1. 布局方式一

最简单的布局方式就是容器顶部是一个 el-header 标签，底部是一个 el-main 标签，如图 2-6 所示。

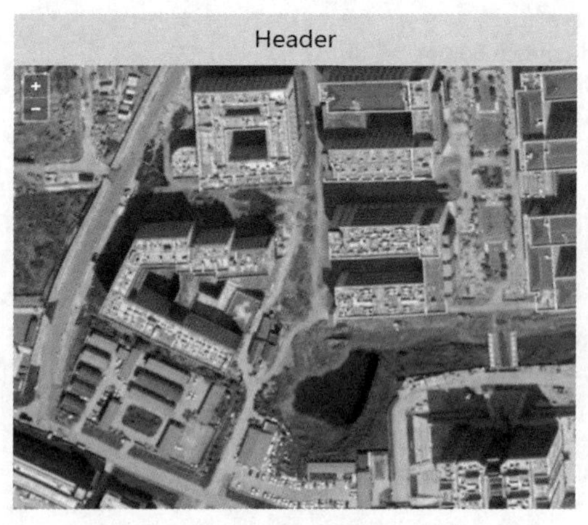

图 2-6　布局方式一

布局方式一的代码如下：

```
<div id="app">
    <!-- 在一个 el-container 标签内包含一个 el-header 和 el-main -->
    <el-container>
        <el-header> Header</el-header>
        <!--将 el-main 标签的 id 赋值为 map，作为地图的容器 -->
```

```
        <el-main id="map"></el-main>
    </el-container>
</div>
```

2. 布局方式二

除了布局方式一，还可以在页面的底部再加上一个 el-footer 标签，如图 2-7 所示。

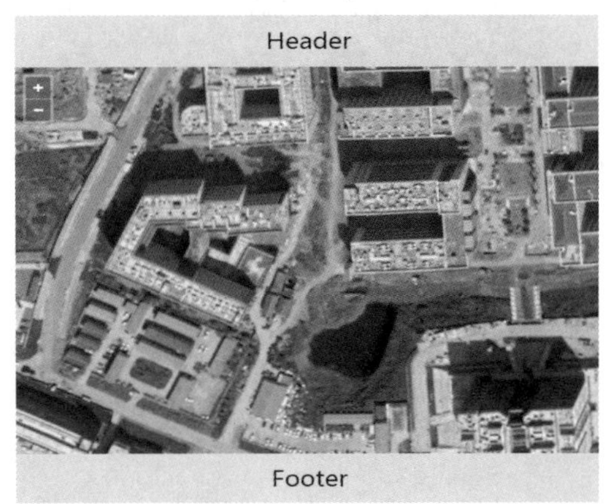

图 2-7　布局方式二

布局方式二的代码如下：

```
<div id="app">
    <!-- 在一个 el-container 标签内包含一个 el-header、一个 el-main 和一个 el-footer -->
    <el-container>
        <el-header>Header</el-header>
        <!--将 el-main 标签的 id 赋值为 map，作为地图的容器 -->
        <el-main id="map"></el-main>
        <el-footer>Footer</el-footer>
    </el-container>
</div>
```

3. 布局方式三

只要 el-container 标签内不包含 el-header 和 el-footer 标签，就可以完成自左向右的布局，如图 2-8 所示。

Element 将 el-aside 标签的宽度默认设置为 300px，读者可以在自己的 CSS 样式中通过"!important"来强制改变该标签的宽度，或者在行内样式中修改标签的宽度。只要自己设置的 CSS 优先级大于默认的 CSS 优先级即可。布局方式三的实现代码如下：

```
<div id="app">
    <!-- 在一个 el-container 标签内包含一个 el-aside 和一个 el-main -->
    <el-container>
        <el-aside>Aside</el-aside>
        <!--将 el-main 标签的 id 赋值为 map，作为地图的容器 -->
        <el-main id="map"></el-main>
```

```
        </el-container>
    </div>
```

图 2-8　布局方式三

4．布局方式四

布局方式四将混合使用由上而下的布局和由左到右的布局，从而得到一个更加完整、漂亮的页面。布局方式四在 el-container 标签中分别放入 el-header、el-aside、el-main、el-footer 标签，如图 2-9 所示。

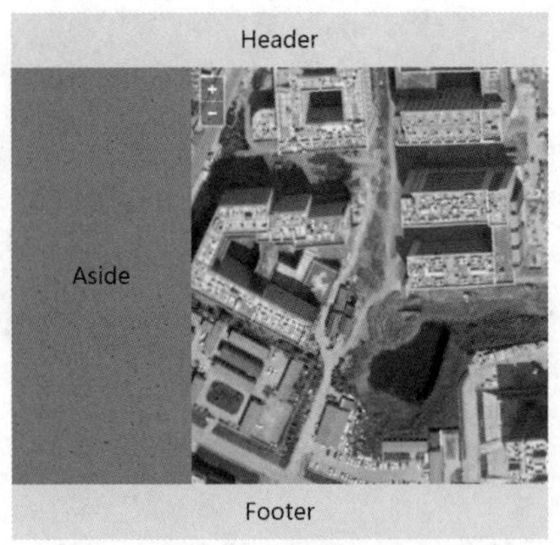

图 2-9　布局方式四

需要注意的是，在布局方式四中，整体布局是由上而下的布局，分为顶部、中部、底部，在中部又进行了由左到右的布局。即 el-container 标签（父 el-container 标签）包含了三部分，顶部为 el-header 标签，中部为 el-container 标签（el-container 子标签），底部为 el-footer 标签，而中部的 el-container 子标签又包含了 el-aside 标签和 el-main 标签。布局方式四的实现代码如下：

```
<div id="app">
    <!-- 在父 el-container 标签内包含一个 el-header 标签、一个 el-container 子标签和一个 el-footer 标签 -->
    <el-container>
```

```
            <el-header>Header</el-header>
            <!-- 在 el-container 子标签内包含了一个 el-aside 标签和一个 el-main 标签 -->
            <el-container>
                <el-aside>Aside</el-aside>
                <!--将 el-main 标签的 id 赋值为 map，作为地图的容器 -->
                <el-main id="map"></el-main>
            </el-container>
            <el-footer>Footer</el-footer>
        </el-container>
    </div>
```

到此，我们就可以使用 el-container、el-header、el-aside、el-main、el-footer 标签进行各种页面布局了。

2.3　Element 的色彩

在色彩方面，Element 提供了一套特定的调色板来指定颜色，带来了不一样的外观视觉感受。Element 的色彩分为主色、辅助色和中性色。主色主要是指鲜艳、友好的蓝色（#409EFF）。辅助色是指用在其他场景的颜色，主要有表示成功的绿色（#67C23A）、表示警告的橙色（#E6A23C）、表示危险的红色（#F56C6C）、表示信息的灰色（#909399）。中性色是指文字颜色、背景颜色、边框颜色等。其中在文字颜色方面，主要分为主要文字（#303133）、常规文字（#606266）、次要文字（#909399）和占位文字（#C0C4CC）；在边框颜色方面，主要分为一级边框（#DCDFE6）、二级边框（#E4E7ED）、三级边框（#EBEEF5）、四级边框（#F2F6FC）；在背景颜色方面，主要分为基础黑色（#000000）、基础白色（#FFFFFF）、透明（Transparent）。在接下来介绍的大部分 Element 标签中，只要涉及背景颜色，都可以通过给标签中的 type 属性赋予 primary、success、info、warning、danger 值来实现不同的背景颜色。

本节将 Element 的色彩与容器布局结合起来，制作一个简单的页面。Element 的色彩效果如图 2-10 所示。

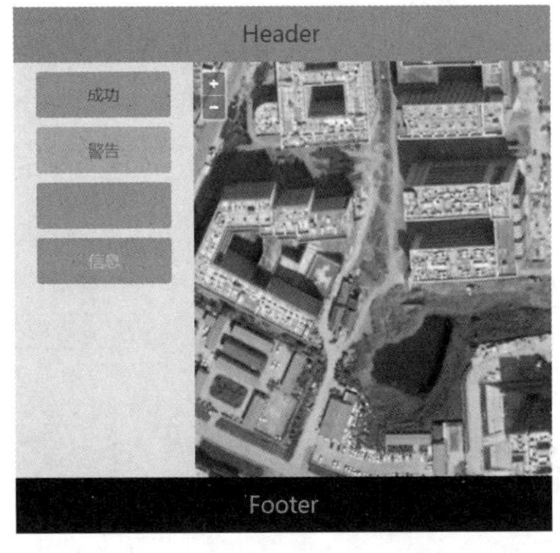

图 2-10　Element 的色彩效果

图 2-10 所示页面的实现代码如下：

```
<style>
    .el-header {
        /*主色*/
        background: #409EFF;
        /*主要文字*/
        color: #303133;
        text-align: center;
        line-height: 60px;
    font-size: 25px;
    }

    .el-aside {
        background: rgb(231, 231, 236);
        /*通过!important 强制改变 el-aside 标签的宽度*/
        width: 200px !important;
    }

    .el-main {
        background-color: rgb(197, 213, 228);
        height: 450px;
        padding: 0 !important;
    }

    .el-footer {
        /*基础黑色*/
        background: #000000;
        /*次要文字*/
        color: #909399;
        text-align: center;
        line-height: 60px;
        font-size: 25px;
    }
    .el-row {
        margin-top: 10px;
    }
    .menue {
        line-height: 50px;
        border-radius: 4px;
        text-align: center;
        font-size: 18px;
    }
    .color1{
        /*表示成功的颜色*/
        background: #67C23A;
        /*表示主要文字的颜色*/
```

```
            color: #303133;
        }
        .color2{
            /*表示警告的颜色*/
            background: #E6A23C;
            /*表示常规文字的颜色*/
            color:#606266;
        }
        .color3{
            /*表示危险的颜色*/
            background: #F56C6C;
            /*表示次要文字的颜色*/
            color: #909399;
        }
        .color4{
            /*表示信息的颜色*/
            background: #909399;
            /*表示占位文字的颜色*/
            color: #C0C4CC;
        }
        </style>
    <div id="app">
        <!-- 在一个 el-container 父标签内包含一个 el-header 标签、一个 el-container 子标签和一个 el-footer
标签 -->
        <el-container>
            <el-header>Header</el-header>
                <!-- 在一个 el-container 子标签内包含了一个 el-aside 标签和一个 el-main 标签 -->
                <el-container>
                    <el-aside>
                        <!-- 在 el-aside 标签中定义三行，每行只有一列，每列跨占 18 栏，并偏移 3 栏 -->
                        <el-row>
                            <el-col :span="18" :offset="3">
                                <div class="menue color1">成功</div>
                            </el-col>
                        </el-row>
                        <el-row>
                            <el-col :span="18" :offset="3">
                                <div class="menue color2">警告</div>
                            </el-col>
                        </el-row>
                        <el-row>
                            <el-col :span="18" :offset="3">
                                <div class="menue color3">危险</div>
                            </el-col>
                        </el-row>
                        <el-row>
                            <el-col :span="18" :offset="3">
```

```
                        <div class="menue color4">信息</div>
                    </el-col>
                </el-row>
            </el-aside>
            <!--将 el-main 标签的 id 赋值为 map，作为地图的容器 -->
            <el-main id="map"></el-main>
        </el-container>
        <el-footer>Footer</el-footer>
    </el-container>
</div>
```

2.4 Element 的字体

Element 在字体方面进行了统一的规范，力求在各种操作系统下都有最佳的效果。Element 提供的字体样式有很多种，具体可在 Element 官网查看，我们可以通过 Element 对字体进行调节。Element 的字体效果如图 2-11 所示。

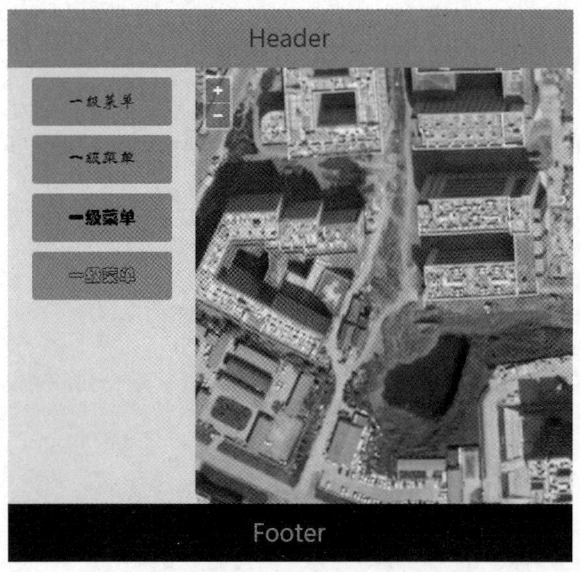

图 2-11 Element 的字体效果

图 2-11 所示页面的实现代码如下：

```
<style>
    .el-header {
        /*主色*/
        background: #409EFF;
        /*主要文字*/
        color: #303133;
        text-align: center;
        line-height: 60px;
```

```
    font-size: 25px;
}

.el-aside {
    background: rgb(231, 231, 236);
    /*通过!important 强制改变 el-aside 标签的宽度*/
    width: 200px !important;
}

.el-main {
    background-color: rgb(197, 213, 228);
    height: 450px;
    padding: 0 !important;
}

.el-footer {
    /*基础黑色*/
    background: #000000;
    /*次要文字*/
    color: #909399;
    text-align: center;
    line-height: 60px;
    font-size: 25px;
}

.el-row {
    margin-top: 10px;
}

.menue {
    line-height: 50px;
    border-radius: 4px;
    text-align: center;
    background: #409EFF;
}

.font1 {
    font-size:large;
    /*华文新魏*/
    font-family: STXinwei;
}

.font2 {
    font-size:large;
    /*华文行楷*/
    font-family: STLiti;
}
```

```
        .font3 {
            font-size:large;
            /*华文隶书*/
            font-family: STHupo;
        }

        .font4 {
            font-size:large;
            /*华文彩云*/
            font-family: STCaiyun;
        }
    </style>
    <div id="app">
        <!-- 在一个 el-container 父标签内包含一个 el-header 标签、一个 el-container 子标签和一个 el-footer
标签 -->
        <el-container>
            <el-header>
                Header
            </el-header>
            <!-- 在一个 el-container 子标签内包含了一个 el-aside 标签和一个 el-main 标签 -->
            <el-container>
                <el-aside>
                    <!-- 在 el-aside 标签中定义三行，每行只有一列，每列跨占 18 栏，并偏移 3 栏 -->
                    <el-row>
                        <el-col :span="18" :offset="3">
                            <div class="menue font1">一级菜单</div>
                        </el-col>
                    </el-row>
                    <el-row>
                        <el-col :span="18" :offset="3">
                            <div class="menue font2">一级菜单</div>
                        </el-col>
                    </el-row>
                    <el-row>
                        <el-col :span="18" :offset="3">
                            <div class="menue font3">一级菜单</div>
                        </el-col>
                    </el-row>
                    <el-row>
                        <el-col :span="18" :offset="3">
                            <div class="menue font4">一级菜单</div>
                        </el-col>
                    </el-row>
                </el-aside>
                <!--将 el-main 标签的 id 赋值为 map，作为地图的容器 -->
                <el-main id="map"></el-main>
```

```
        </el-container>
        <el-footer>Footer</el-footer>
    </el-container>
</div>
```

2.5　Element 的边框

Element 提供了多种边框的圆角样式。Element 的边框效果如图 2-12 所示。

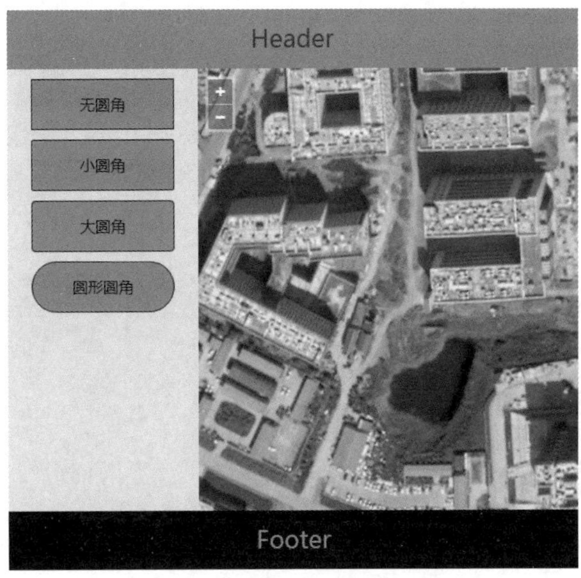

图 2-12　Element 的边框效果

图 2-12 所示页面的实现代码如下：

```
<style>
    .el-header {
        /*主色*/
        background: #409EFF;
        /*主要文字*/
        color: #303133;
        text-align: center;
        line-height: 60px;
        font-size: 25px;
    }

    .el-aside {
        background: rgb(231, 231, 236);
        /*通过!important 强制改变 el-aside 标签的宽度*/
        width: 200px !important;
    }
```

```
.el-main {
    background-color: rgb(197, 213, 228);
    height: 450px;
    padding: 0 !important;
}

.el-footer {
    /*基础黑色*/
    background: #000000;
    /*次要文字*/
    color: #909399;
    text-align: center;
    line-height: 60px;
    font-size: 25px;
}

.el-row {
    margin-top: 10px;
}

.menue {
    line-height: 50px;
    border-radius: 4px;
    text-align: center;
    background: #409EFF;
}

.border1 {
    border: 1px solid black;
    /*无圆角*/
    border-radius: 0px;
}

.border2 {
    border: 1px solid black;
    /*小圆角*/
    border-radius: 2px;
}

.border3 {
    border: 1px solid black;
    /*大圆角*/
    border-radius: 4px;
}

.border4 {
```

```
            border: 1px solid black;
            /*圆形圆角*/
            border-radius: 30px;
        }
    </style>
    <div id="app">
        <!-- 在一个 el-container 父标签内包含一个 el-header 标签、一个 el-container 子标签和一个 el-footer
标签 -->
        <el-container>
            <el-header>
                Header
            </el-header>
            <!-- 在一个 el-container 子标签内包含了一个 el-aside 标签和一个 el-main 标签 -->
            <el-container>
                <el-aside>
                    <!-- 在 el-aside 标签中定义三行，每行只有一列，每列跨占 18 栏，并偏移 3 栏 -->
                    <el-row>
                        <el-col :span="18" :offset="3">
                            <div class="menue border1">无圆角</div>
                        </el-col>
                    </el-row>
                    <el-row>
                        <el-col :span="18" :offset="3">
                            <div class="menue border2">小圆角</div>
                        </el-col>
                    </el-row>
                    <el-row>
                        <el-col :span="18" :offset="3">
                            <div class="menue border3">大圆角</div>
                        </el-col>
                    </el-row>
                    <el-row>
                        <el-col :span="18" :offset="3">
                            <div class="menue border4">圆形圆角</div>
                        </el-col>
                    </el-row>
                </el-aside>
                <!--将 el-main 标签的 id 赋值为 map，作为地图的容器 -->
                <el-main id="map"></el-main>
            </el-container>
            <el-footer>Footer</el-footer>
        </el-container>
    </div>
```

2.6 Element 的图标

在前端开发中，图标是必不可少的部分。精美的图标不仅可以提高开发效率，还可以使页面更加美观。阿里巴巴矢量图标库包含了大量、丰富的图标，可以满足前端开发的需求。Element 提供了一套常用的图标集合，可以通过类 el-icon-iconName 来使用这些图标。

在使用 Element 的图标之前，需要先在 Element 官网下载 element-icons.ttf 和 element-icons.woff 文件。具体方法为：在 Element 官网的首页中单击“组件”按钮，然后单击“unpkg.com/element-ui”，在弹出的页面中打开文件夹“lib\theme-chalk\fonts”，可看到文件 element-icons.ttf 和文件 element-icons.woff，如图 2-13 所示。

UNPKG

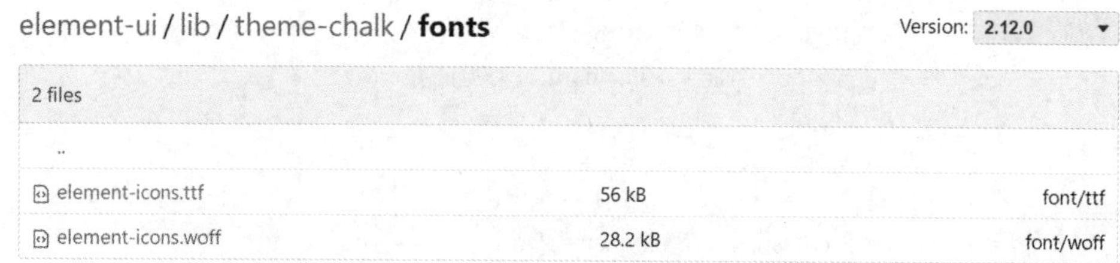

图 2-13　文件 element-icons.ttf 和文件 element-icons.woff

单击文件 element-icons.ttf，可打开该文件的页面，如图 2-14 所示，单击图中的“View Raw”即可下载该文件。使用同样的方法下载文件 element-icons.woff。

UNPKG

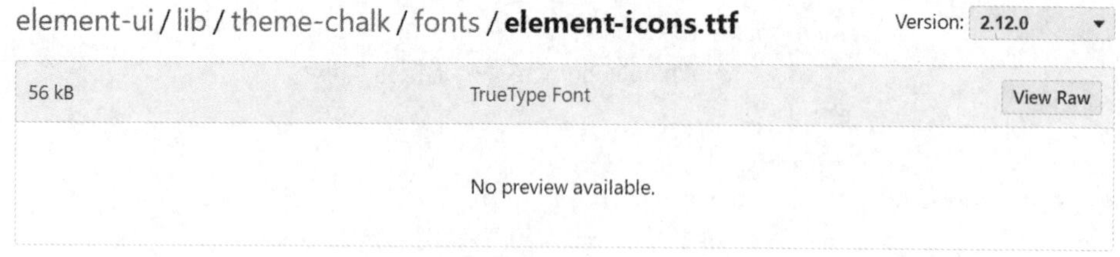

图 2-14　文件 element-icons.ttf 的页面

在项目的 Element 目录下新建文件夹 fonts（见图 2-15），然后将下载后的两个文件 element-icons.ttf 和 element-icons.woff 复制到文件夹 fonts 中即可。

至此，我们就可以使用 Element 的图标了。下面将通过一个简单的页面（见图 2-16）来介绍 Element 图标的使用方法。

图 2-15　新建的文件夹 fonts

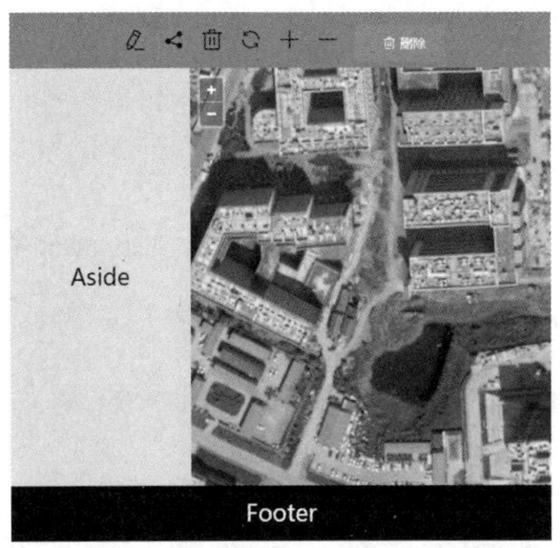

图 2-16　使用 Element 图标的页面

图 2-16 所示的页面只展示了部分图标，要想使用更多的 Element 图标，请查看 Element 官网首页的“组件”下的“Icon 图标”，如图 2-17 所示。

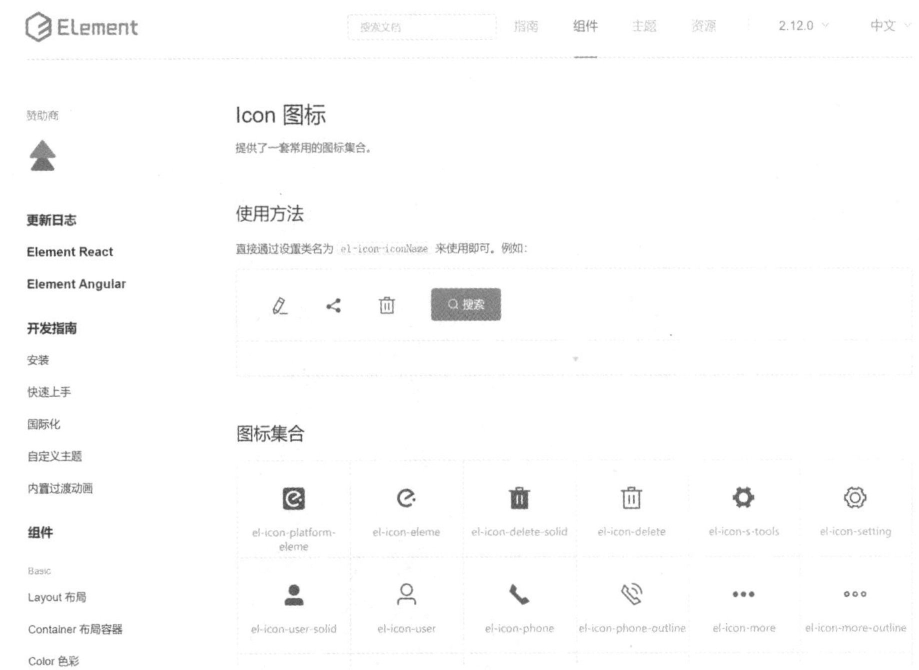

图 2-17　Element 官网中“组件”下的“Icon 图标”

在使用标签时，通过为标签 i 定义不同的 class 值可以获取不同类型的图标；在使用按钮时，通过定义 icon 属性的值可以将图标融入按钮。实现代码如下：

```
<div id="app">
    <el-container>
        <!--通过为 i 标签定义不同的 class 值可获取不同类型的图标；在使用按钮时，通过定义 icon
属性的值可以将图标融入按钮-->
            <el-header>
                <i class="el-icon-edit "></i>
                <i class="el-icon-share "></i>
                <i class="el-icon-delete "></i>
                <i class="el-icon-refresh "></i>
                <i class="el-icon-plus"></i>
                <i class="el-icon-minus"></i>
            <el-button type="danger" icon="el-icon-delete" class="btn">删除</el-button>
            </el-header>
            <el-container>
            <el-aside>Aside</el-aside>
            <!--将 el-main 标签的 id 赋值为 map，作为地图的容器 -->
            <el-main id="map"></el-main>
            </el-container>
        <el-footer>Footer</el-footer>
    </el-container>
</div>
```

2.7 Element 的按钮

2.7.1 按钮的样式定义

Element 为按钮提供了不同的样式与用法，本节将从基础用法、图标按钮、文字按钮、禁用状态、加载中、不同尺寸、按钮组等方面讲解按钮的样式与用法。本节首先结合地图展示按钮的样式，如图 2-18 所示；然后对按钮的样式与用法进行讲解。

图 2-18　结合地图展示按钮的样式

在创建按钮样式前需要先创建 Vue 实例，由于本节的示例比较简单，因此 data 和 methods 都为空。

```
var vm = new Vue({
    el: "#app",
    data: {
    },
    methods:{
    }
})
```

（1）基础用法。Element 通过对 el-button 标签的封装来提供按钮，读者可以通过在 el-button 标签中设置 type、plain、round 和 circle 等属性来丰富按钮的样式。

```
<el-row>
    <el-button>默认按钮</el-button>
    <el-button type="primary">主要按钮</el-button>
</el-row>
<el-row>
    <el-button round>圆角按钮</el-button>
    <el-button type="info" round>信息按钮</el-button>
</el-row>
```

（2）图标按钮。图标按钮将图标添加到了按钮中，只需要对 el-button 标签中的 icon 属性赋予不同图标名字即可实现图标按钮。代码如下：

```
<el-button icon="el-icon-search"></el-button>
<el-button type="primary" icon="el-icon-edit" circle></el-button>
<el-button type="success" icon="el-icon-check" circle></el-button>
<el-button type="info" icon="el-icon-message" circle></el-button>
```

（3）文字按钮。顾名思义，文字按钮将文字添加到了按钮中，只需要对 el-button 标签中的 type 属性赋予 text 值即可实现文字按钮。代码如下：

```
<el-button type="text">文字按钮</el-button>
```

（4）禁用状态。禁用状态是指按钮不能被单击，将 el-button 标签中 disabled 属性的值设为 true 即可使按钮处于禁用状态。代码如下：

```
<el-button plain disabled>按钮一</el-button>
<el-button type="success" disabled>按钮二</el-button>
```

（5）加载中。加载中的按钮是指单击按钮后，数据还处于加载状态。只需要在 el-button 中添加 loading 属性并赋予 true 值即可实现加载中。代码如下：

```
<el-button type="primary" :loading="true">加载中</el-button>
```

（6）不同尺寸。不同尺寸是指对按钮的大小进行了规划。通过对 el-button 标签中的 size 属性赋予 medium、small、mini 等属性即可实现中等按钮、小型按钮、超小按钮。代码如下：

```
<el-button>默认按钮</el-button>
<el-button size="medium">中等按钮</el-button>
<el-button size="small">小型按钮</el-button>
<el-button size="mini">超小按钮</el-button>
```

（7）按钮组。按钮组是指多个按钮在一起出现，通过使用 el-button-group 标签来嵌入按钮即可实现按钮组。代码如下：

```
<el-button-group>
    <el-button type="primary" icon="el-icon-arrow-left">上一页</el-button>
    <el-button type="primary">下一页<i class="el-icon-arrow-right el-icon--right"></i></el-button>
</el-button-group>
```

2.7.2 按钮的事件绑定

本节主要介绍为页面按钮绑定事件的方法，图 2-19 所示的页面布局了三个按钮，分别是“警告按钮”按钮、“地图放大”按钮和“地图缩小”按钮。

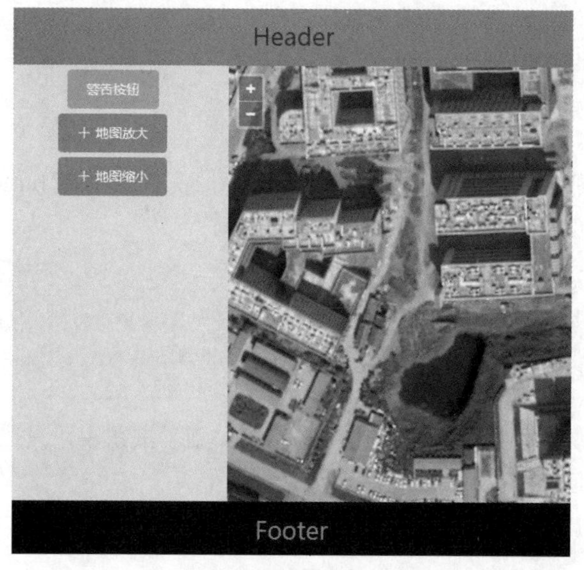

图 2-19　布局了三个按钮的页面

单击图 2-19 中的“警告按钮”按钮可弹出提示框，如图 2-20 所示；单击“地图放大”按钮可放大地图，如图 2-21 所示；单击“地图缩小”按钮可缩小地图，如图 2-22 所示。

为上述三个按钮绑定事件的方法是：创建 Vue 实例，将 methods 中的 warningBtnHandleClick 作为单击“警告按钮”按钮时的响应事件，将 methods 中的 zoomUpBtnHandleClick 作为单击“地图放大”按钮时的响应事件，将 methods 中的 zoomOutBtnHandleClick 作为单击“地图缩小”按钮时的响应事件。实现代码如下：

图 2-20　单击“警告按钮”按钮的效果

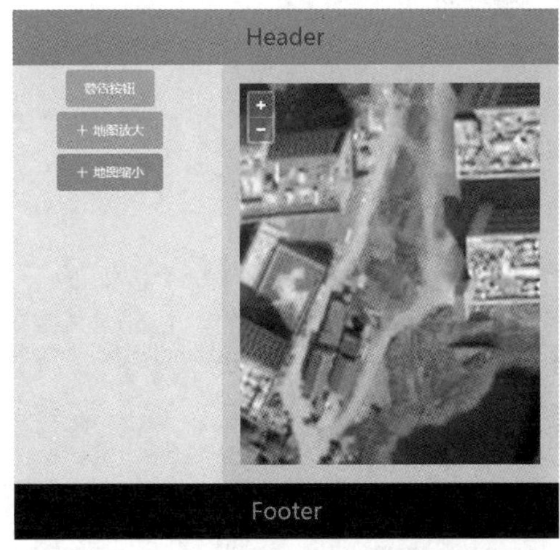

图 2-21　单击“地图放大”按钮的效果

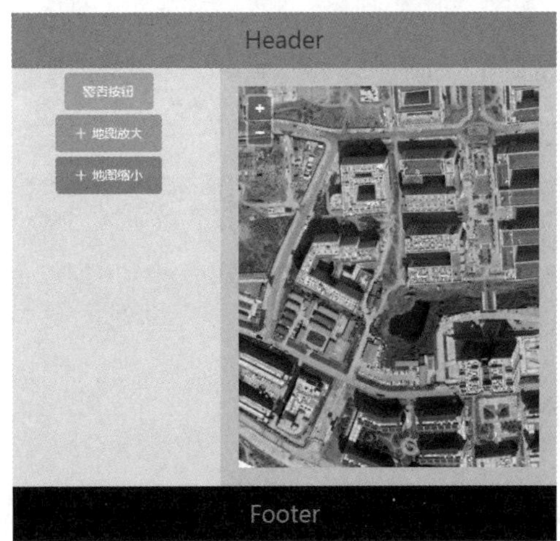

图 2-22　单击“地图缩小”按钮的效果

```
var vm = new Vue({
    el: "#app",
    //定义事件
    methods: {
        //单击“警告按钮”按钮时的事件
        warningBtnHandleClick: function () {
            alert("警告!");
        },
        //单击“地图放大”按钮时的事件
        zoomUpBtnHandleClick: function () {
            var zoom = map.getView().getZoom();
            map.getView().setZoom(zoom + 1);
        },
        //单击“地图缩小”按钮时的事件
        zoomOutBtnHandleClick: function () {
            var zoom = map.getView().getZoom();
            map.getView().setZoom(zoom - 1);
        },

    }
})
```

页面布局代码如下：

```
<el-row>
    <el-col :offset="6" :span="12">
        <el-button type="warning" @click="warningBtnHandleClick">警告按钮</el-button>
    </el-col>
</el-row>
<el-row>
    <el-col :offset="5" :span="12">
        <el-button type="primary" icon="el-icon-plus" @click="zoomUpBtnHandleClick">地图放大
</el-button>
    </el-col>
</el-row>
<el-row>
    <el-col :offset="5" :span="12">
        <el-button type="primary" icon="el-icon-plus" @click="zoomOutBtnHandleClick">地图缩小
        </el-button>
    </el-col>
</el-row>
```

2.8 Element 的文字链接

Element 为文字链接提供了不同的样式与用法，本节将从基础用法、下画线、图标、禁用状态等方面介绍文字链接的样式与用法。本节首先结合地图展示文字链接的样式，如图 2-23

所示；然后对文字链接的样式与用法进行讲解。

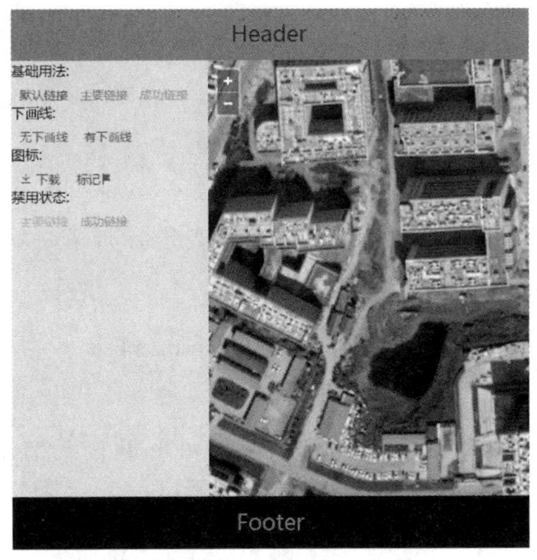

图 2-23　文字链接的样式

单击“基础用法”中的“默认链接”，会跳转到 Vue 官网首页，如图 2-24 所示。

图 2-24　Vue 官网的首页

创建 Vue 实例，由于本示例比较简单，因此 data 和 methods 都为空。

```
var vm = new Vue({
    el: "#app",
    data: {
    },
    methods:{
    }
})
```

（1）基础用法。Element 通过对 el-link 标签的封装提供了文字链接，href 属性用来标注文字链接的网址；可通过将 type 属性设置为 primary、success、warning、danger 和 info 来定义不同的颜色。代码如下：

```
<!-- href 属性所对应的值是指某个网站的网址，若 target 的值为空，则表示在原网页上打开一个新的网页；若 target 的值为 blank，则会在一个新的窗口中打开网页 -->
<el-link href="https://cn.vuejs.org/" target="blank">默认链接</el-link>  
<el-link type="primary">主要链接</el-link>  
<el-link type="success">成功链接</el-link>
```

（2）下画线。underline 属性的值默认是 true，表示具有下画线。若将 underline 属性的值设置为 false，则下画线会消失。代码如下：

```
<el-link :underline="false">无下画线</el-link>   
<el-link>有下画线</el-link>
```

（3）图标。既可以在 el-link 标签中添加并设置 icon 属性的值来使用图标，此时图标在文字前面；也可以在 el-link 标签中添加 i 标签来使用图标，此时图标在文字后面。代码如下：

```
<el-link icon="el-icon-download">下载</el-link>   
<el-link>标注<i class="el-icon-s-flag"></i> </el-link>
```

（4）禁用状态。在 el-link 标签中添加 disabled 属性，可以使文字链接处于禁用状态。代码如下：

```
<el-link type="primary" disabled>主要链接</el-link>  
<el-link type="success">成功链接</el-link>
```

2.9 思考与练习题

（1）参考 2.1 节和 2.2 节，结合所学的标签来搭建一个布局容器并在该布局容器中添加布局。

（2）参考 2.6 节和 2.7 节来定义带有标签的按钮，并在单击该按钮时弹出一个对话框。

（3）参考 2.6 节和 2.7 节来定义带有标签的文字链接，并在单击该标签时跳转到 OpenLayers 的官网。

第 3 章 表单组件

Element 为表单提供了非常丰富、详细的样式，本章结合 Vue 和 OpenLayers 来详细介绍 Element 的各个表单组件。

3.1 单选框（Radio）

Element 为单选框提供了不同的样式与用法，本节将从基础用法、单选框组、按钮样式、带有边框、禁用状态等方面讲解单选框的样式与用法。本节首先结合地图展示单选框的样式，如图 3-1 所示；然后对单选框的样式与用法进行讲解。

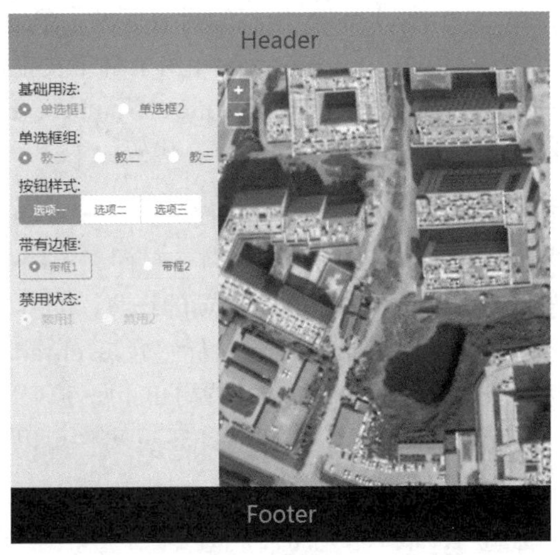

图 3-1　单选框样式

本示例首先创建 Vue 实例，其中 data 中的 radio1 到 radio5 分别是各个 el-radio-group 标签中 v-model 属性所绑定的变量；在单选框的选中值改变时，methods 中的 handleChange 事件会被触发，回调参数为目前的选中值，该值与 el-radio 标签中 label 属性所对应的值联动；然后结合 OpenLayers 实现地图视角的改变。代码如下：

```
var vm = new Vue({
    el: "#app",
    data: {
        radio1: 1,
        radio2: 1,
        radio3: "选项一",
        radio4: 1,
        radio5: 1
    },
    methods: {
        //在单选框的选中值改变时的响应事件
        handleChange: function (val) {
            if (this.radio2 === 1) {
                map.getView().setCenter([2700, 1600]);
            }
            if (this.radio2 === 2) {
                map.getView().setCenter([3000, 2000]);
            }
            else if (this.radio2 === 3) {
                map.getView().setCenter([3500, 1200]);
            }
        }
    }
})
```

（1）基础用法。Element 通过对 el-radio 标签的封装提供了单选框，该标签中的 v-model 和 label 属性完成了单选框的基础用法，其中 v-model 用来绑定变量，选中后该变量的值为相应的 label 属性的值，label 属性的值类型可以是 String、Number、Boolean，另外若标签中的内容为空，则 label 所对应的属性值就是标签的内容。代码如下：

```
<el-radio v-model="radio1" :label="1">单选框 1</el-radio>
<el-radio v-model="radio1" :label="2">单选框 2</el-radio>
```

（2）单选框组。单选框组适用于有多个互斥选项的场景，当存在多个单选框时，使用单选框组会更加方便。单选框组是通过 el-radio-group 标签包裹 el-radio 标签来完成的，可以在 el-radio-group 标签中绑定 v-model，在 el-radio 中设置 label 属性，无须为每个 el-radio 标签都绑定 v-model，另外，在 el-radio-group 标签中绑定 change 事件可以监听单选框组的变化，并传入一个参数 value。代码如下：

```
<el-radio-group v-model="radio2" @change="handleChange">
    <el-radio :label="1">教一</el-radio>
    <el-radio :label="2">教二</el-radio>
    <el-radio :label="3">教三</el-radio>
</el-radio-group>
```

（3）按钮样式。按钮样式适合于选项卡的场景，只需要将 el-radio 标签改为 el-radio-button 标签即可实现按钮样式。另外，Element 为 el-radio-group 标签提供的 size 属性可以用来调整按钮的大小，size 属性的值可以是 medium、small 和 mini。代码如下：

```
<el-radio-group v-model="radio3" size=small>
    <el-radio-button label="选项一"></el-radio-button>
    <el-radio-button label="选项二"></el-radio-button>
    <el-radio-button label="选项三"></el-radio-button>
</el-radio-group>
```

（4）带有边框。若想为单选框加上边框，只需要为 el-radio 标签添加 border 属性即可。代码如下：

```
<el-radio v-model="radio4" :label="1" border size=mini>带框 1</el-radio>
<el-radio v-model="radio4" :label="2" border size=mini>带框 2</el-radio>
```

（5）禁用状态。单选框的禁用状态和按钮的禁用状态一样，只需要在标签中将 disabled 属性设置为 true 即可。代码如下：

```
<el-radio-group v-model="radio5" :disabled="true" >
    <el-radio :label="1">禁用 1</el-radio>
    <el-radio :label="2">禁用 2</el-radio>
</el-radio-group>
```

3.2 多选框（Checkbox）

Element 为多选框提供了不同的样式与用法，本节将从基础用法、多选框禁用状态、多选框组、indeterminate 状态、可选项目数量限制、按钮样式、带有边框等方面讲解多选框的样式与用法。本节首先结合地图展示多选框的样式，如图 3-2 所示；然后对多选框的样式与用法进行讲解。

单击“带有边框”中第二个单选框，即“框 2”按钮，会弹出一个提示框，如图 3-3 所示。

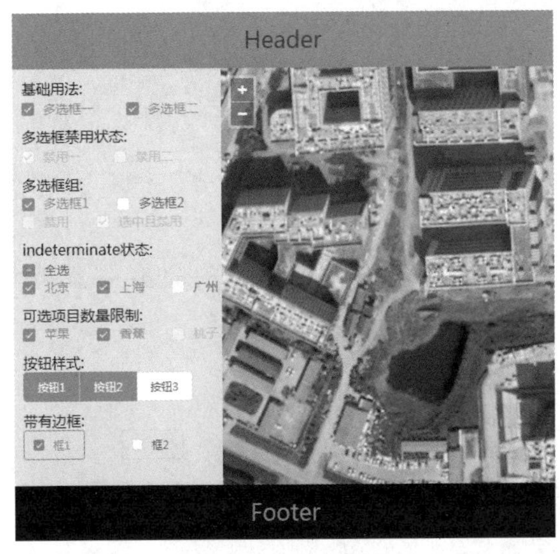

图 3-2　多选框样式

图 3-3　弹出提示框的效果

本示例首先创建 Vue 实例，其中 data 中的各个变量与 el-checkbox-group 标签中的 v-model 属性和 el-checkbox 标签中的 label 属性等进行联动，methods 中的 handleCheckAllChange 在 intermediate 状态中全选框的状态改变时被触发，handleCheckCitysChange 在 intermediate 状态中多选框状态改变时被触发，handleCheckBorder2Change 在带有边框中多选框状态改变时被触发；然后结合 OpenLayers 实现地图视角的改变。代码如下：

```
var vm = new Vue({
    el: "#app",
    data: {
        //1.基础用法所对应的变量
        checked1_1: true,
        checked1_2: true,
        //2.多选框禁用状态所对应的变量
        checked2_1: true,
        checked2_2: false,
        //3.多选框组所对应的变量
        checked3: ["多选框 1", "选中且禁用"],
        //4.indeterminate 状态所对应的变量
        isIndeterminate: true,
        cityList: ["北京", "上海", "广州"],
        checkAll: false,
        checkCitys: ["北京", "上海"],
        //5.可选项目数量限制所对应的变量
        checkFruits: ["苹果", "香蕉"],
        fruits: ["苹果", "香蕉", "桃子"],
        //6.按钮样式所对应的变量
        checkBtns: ["按钮 1", "按钮 2"],
        btns: ["按钮 1", "按钮 2", "按钮 3"],
        //7.带有边框所对应的变量
        checkBorder: ["框 1"]
    },
    methods: {
        //在 intermediate 状态中全选框的状态改变时的响应事件
        handleCheckAllChange: function (val) {
            /*参数 val 为 true 或 false。若勾选则为 true，将 checkCitys 赋值为 cityList；若不勾选则为
false，将 checkCitys 赋值为空数组*/
            this.checkCitys = val ? this.cityList : [];
            /*该事件执行后，各个多选框要么为全选状态，要么为空状态，可以将 indeterminate 赋值
为 false*/
            this.isIndeterminate = false;
        },
        //在 intermediate 状态中多选框的状态改变时的响应事件
        handleCheckCitysChange: function (val) {
            /*若所有的多选框都被勾选，则将 checkAll 框赋值为 true，否则赋值为 false*/
            this.checkAll = val.length === this.cityList.length;
            /*参数 val 传递的是一个数组，只有当部分多选框被勾选时，indeterminate 状态才能被激
```

```
活，才能被赋值为 true*/
                this.isIndeterminate = val.length > 0 && val.length < this.cityList.length;
            },
            //在带边框样式中多选框的状态改变时的响应事件
            handleCheckBorder2Change: function () {
                alert("框 2");
            }
        }
    })
```

（1）基础用法。Element 通过 el-checkbox 标签对多选框进行了封装，该标签通过 v-model 属性来绑定变量，若变量的值是 true，则表示被选中；若变量的值是 false，则表示未被选中。代码如下：

```
<el-checkbox v-model="checked1_1">多选框一</el-checkbox>
<el-checkbox v-model="checked1_2">多选框二</el-checkbox>
```

（2）多选框禁用状态。只需要在 el-checkbox 标签中添加 disabled 属性即可实现多选框禁用状态，多选框禁用状态可分为未选中禁用状态和已选中禁用状态。代码如下：

```
<el-checkbox v-model="checked2_1" disabled>禁用一</el-checkbox>
<el-checkbox v-model="checked2_2" disabled>禁用二</el-checkbox>
```

（3）多选框组。与单选框组类似，多选框组同样是为了把多个元素合并为一组进行管理。通过 el-checkbox-group 标签与多个子元素 el-checkbox 标签的结合可实现多选框组。在 el-checkbox-group 标签中使用 v-model 绑定一个数组，将该数组的元素与各 el-checkbox 标签中的 label 属性的值进行匹配，可以决定各个多选框的选中状态。代码如下：

```
<el-checkbox-group v-model="checked3">
    <el-checkbox label="多选框 1"></el-checkbox>
    <el-checkbox label="多选框 2"></el-checkbox>
    <el-checkbox label="禁用" disabled></el-checkbox>
    <el-checkbox label="选中且禁用" disabled></el-checkbox>
</el-checkbox-group>
```

（4）indeterminate 状态。indeterminate 状态表示 checkbox 的不确定状态，一般用于实现全选的效果。indeterminate 状态的使用方法为：首先定义一个 el-checkbox 标签，在该标签上设置两个属性和绑定一个事件；然后通过给 indeterminate 属性绑定一个变量用于说明该标签是 indeterminate 状态，通过 v-model 绑定一个变量用于控制是否全选，通过绑定 change 来监听该标签的变化；最后定义一个多选框组，通过 Vue 提供的 v-for 对多个多选框进行快速的列表渲染。代码如下：

```
<el-checkbox
    :indeterminate="isIndeterminate"
    v-model="checkAll"
    @change="handleCheckAllChange">全选</el-checkbox>
    <el-checkbox-group v-model="checkCitys" @change="handleCheckCitysChange">
    <!--  注意 label 前这个冒号不能够省略。因为 city 是个变量，而上面示例中的 label 对应的是常量，
```

```
故不需要带冒号 -->
        <el-checkbox v-for="city of cityList" :label="city">{{city}}</el-checkbox>
    </el-checkbox-group>
```

（5）可选项目数量限制。在某些情况下，需要对多选框的最少选中状态和最多选中状态进行限制。只需要在 el-checkbox-group 标签中设置 min 和 max 属性即可实现可选项目数量限制。

（6）按钮样式。与单选框的按钮样式类似，Element 同样为多选框提供了按钮样式，只需要将 el-checkbox 改为 el-checkbox-group 即可实现多选框的按钮样式。代码如下：

```
    <el-checkbox-group v-model="checkFruits" :min="1" :max="2">
        <el-checkbox v-for="fruit of fruits" :label="fruit">{{fruit}}</el-checkbox>
    </el-checkbox-group>
```

（7）带有边框。与单选框的带有边框类似，只需要在 el-checkbox 标签中设置 border 属性即可实现多选框的带有边框。代码如下：

```
    <el-checkbox-group v-model="checkBtns" size="small">
        <el-checkbox-button
        v-for="btn of btns" :label="btn">{{btn}}</el-checkbox-button>
    </el-checkbox-group>
```

3.3 输入框（Input）

Element 为输入框提供了不同的样式与用法，本节将从基础用法、禁用状态、可清空、密码框、带 icon 的输入框用法、文本域、可自适应文本高度的文本域、尺寸、复合型输入框、带输入建议、自定义模板、输入长度限制等方面讲解输入框的样式与用法。

由于输入框的样式比较丰富，因此本节首先对两个页面进行布局，然后结合地图展示输入框的样式，如图 3-4 和图 3-5 所示，最后对输入框的样式与用法进行讲解。

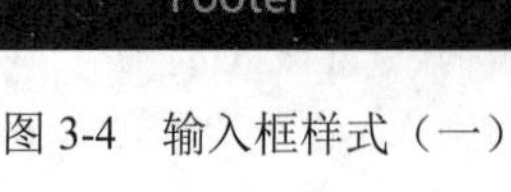
图 3-4　输入框样式（一）

图 3-5　输入框样式（二）

在第一个页面中（见图 3-4），首先创建 Vue 实例，其中 data 中的 input1 至 input7 是各个 el-input 标签中 v-model 属性所绑定的变量；methods 中的 handleClick 事件在单击“查询”按钮时被触发，此时可获取“基础用法”中输入的内容，然后结合 OpenLayers 实现地图视角的改变。代码如下：

```
var vm = new Vue({
    el: "#app",
    data: {
        input1: "",
        input2: "",
        input3: "",
        input4: "",
        input5_1: "",
        input5_2: "",
        input6: "",
        input7: ""
    },
    methods: {
        handleClick: function () {
            if (this.input1 === "一号教学楼") {
                map.getView().setCenter([3000, 1000]);
            }
        }
    }
})
```

在第二个页面中（见图 3-5），创建 Vue 实例，其中 data 中的各属性与 el-input 标签中 v-model 属性所绑定的变量联动；methods 中的 querySearch 是返回输入建议的方法。代码如下：

```
var vm = new Vue({
    el: "#app",
    data: {
        input8:"",
        input9: "",
        sites: [{ "value": "食堂一", "address": "A 点" },
        { "value": "食堂二", "address": "B 点" },
        { "value": "图书馆", "address": "C 点" },
        { "value": "教学楼", "address": "D 点" },
        { "value": "体育场", "address": "E 点" },
        { "value": "办公楼", "address": "F 点" },],
        state1: '',
        state2: '',
        state3: '',
        input13: '',
    },
    methods: {
        querySearch(queryString, cb) {
```

```
                //如果输入框的信息为空，那么 autocomplete 模板就是所有的数据
                if (queryString === "") {
                    results = this.sites;
                }
                /*若输入框的信息不为空，则逐个筛选所有地点，并将包含输入框内容的地点返回到
autocomplete 模板*/
                else {
                    var newres = this.sites.filter(function (result) {
                        return result.value.indexOf(queryString) === 0;
                    })
                results = newres;
                }
                //调用 callback 返回建议列表中的数据
                cb(results);
            }
        },
    })
```

（1）基础用法。Element 通过对 el-input 标签的封装提供了输入框，通过 v-model 属性对输入框的值进行监听，通过 placeholder 属性对输入框的占位符进行绑定。代码如下：

```
<el-input v-model="input1" placeholder="请输入内容" size="mini"></el-input>
```

（2）禁用状态。Element 通过对 el-input 标签添加 disabled 属性来控制输入框的禁用状态。代码如下：

```
<el-input v-model="input2" placeholder="请输入内容" disabled size="mini"></el-input>
```

（3）可清空。Element 通过对 el-input 标签添加 clearable 属性来控制输入框是否可清空。代码如下：

```
<el-input v-model="input3" placeholder="请输入内容" size="mini" clearable></el-input>
```

（4）密码框。Element 通过对 el-input 标签添加 show-password 属性来控制输入框是否为密码框。代码如下：

```
<el-input v-model="input4" placeholder="请输入内容" size="mini" show-password></el-input>
```

（5）带 icon 的输入框。Element 在封装输入框时，提供了带图标的输入框样式。其中 Element 提供了两种方式，分别为属性方式和 slot 方式。属性方式是通过在 el-input 标签添加 suffix-icon 属性和 prefix-icon 属性实现的，其中 suffix-icon 属性表示图标在输入框的尾部，prefix-icon 属性表示图标在输入框的首部。slot 方式是通过在 input 标签中添加 i 标签来实现的，i 标签中的 slot 属性值可以是 suffix 和 prefix，suffix 表示图标在输入框的尾部，prefix 表示图标在输入框的首部。代码如下：

```
<el-input v-model="input5_1" placeholder="请输入内容"
size="mini" suffix-icon="el-icon-search"> </el-input>
<el-input v-model="input5_2" placeholder="请输入内容" size="mini">
    <i slot="prefix" class="el-input__icon el-icon-search"></i>
```

```
</el-input>
```

（6）文本域。文本域用于在输入框中输入多行文本信息，可通过在 el-input 标签中添加 type 属性并赋予 textarea 值来实现文本域，可通过 rows 属性来控制文本域的高度。代码如下：

```
<el-input v-model="input6" placeholder="请输入内容"
type="textarea" :rows="2"></el-input>
```

（7）可自适应文本高度的文本域。在文本域中，可通过 autosize 属性使文本域的高度根据文本高度进行自动调整，并且 autosize 属性还可以设置一个对象来指定最小行数和最大行数。代码如下：

```
<el-input v-model="input7" placeholder="请输入内容" type="textarea"
:autosize="{ minRows: 2, maxRows: 4}"></el-input>
```

（8）尺寸。在 el-input 标签中添加 size 属性可以控制输入框的尺寸，尺寸共分为 large、medium、small、mini 四种，默认的是 large。代码如下：

```
<el-input v-model="input1" placeholder="请输入内容" size="medium"></el-input>
<el-input v-model="input1" placeholder="请输入内容" size="small"></el-input>
<el-input v-model="input1" placeholder="请输入内容" size="mini"></el-input>
```

（9）复合型输入框。复合型输入框可以在输入框中添加其他元素，元素一般为标签、按钮和下拉框，通过 slot 属性可以控制元素在输入框中的位置。代码如下：

```
<el-input placeholder="请输入内容" v-model="input9" class="input-with-select">
<el-button slot="append" icon="el-icon-search"></el-button>
```

（10）带输入建议。Element 通过 el-autocomplete 标签提供了带输入建议的输入框。带输入建议的输入框可以智能地推荐用户想要输入的信息。带输入建议的输入框有两种，分别是列表型输入框和输入匹配型输入框。与列表型输入框相比，输入匹配型输入框将 el-autocomplete 标签中的 trigger-on-focus 属性设置为 flase，该属性默认为 true。el-autocomplete 标签中的 fetch-suggestions 属性是一个方法属性，用于返回输入的建议，如“querySearch (queryString, cb)”，queryString 是输入框的内容，cb 是个回调函数，可以将建议的数据通过 cb(data)返回到 autocomplete 模板中（autocomplete 模板是指带智能提示的下拉框）。代码如下：

```
<el-autocomplete v-model="state1" :fetch-suggestions="querySearch" placeholder="请输入内容"></el-autocomplete>
<el-autocomplete v-model="state2" :fetch-suggestions="querySearch" placeholder="请输入内容":trigger-on-
focus="false"></el-autocomplete>
```

（11）自定义模板。自定义模板可以在带输入建议的输入框中添加更加复杂的样式，如添加图标。通过 Vue.js 的作用域插槽 scoped slot 可以自定义输入建议的模板，scope 的参数为 item，表示当前输入建议对象。代码如下：

```
<el-autocomplete v-model="state3" :fetch-suggestions="querySearch" placeholder="请输入内容">
    <i class="el-icon-edit el-input__icon" slot="suffix"></i>
  <template slot-scope="{ item }">
      <div class="name">{{ item.value }}</div>
      <span class="addr">{{ item.address }}</span>
```

```
        </template>
    </el-autocomplete>
```

（12）输入长度限制。Element 为 el-input 标签提供了 maxlength 和 minlength 属性，可用于限制输入框的最大输入长度和最小输入长度。对于类型为 text 或 textarea 的输入框，在使用 maxlength 属性限制最大输入长度时，可通过设置 show-word-limit 属性来统计字数。代码如下：

```
    <el-input type="text" placeholder="请输入内容" v-model="input13" maxlength="10" show-word-limit>
</el-input>
```

3.4 计数器（InputNumber）

Element 为计数器提供了不同的样式与用法，本节将从基础用法、禁用状态、步数、严格步数、精度、尺寸、按钮位置等方面讲解计数器的样式与用法。本节首先结合地图展示计数器的样式，如图 3-6 所示；然后对计数器的样式与用法进行讲解。

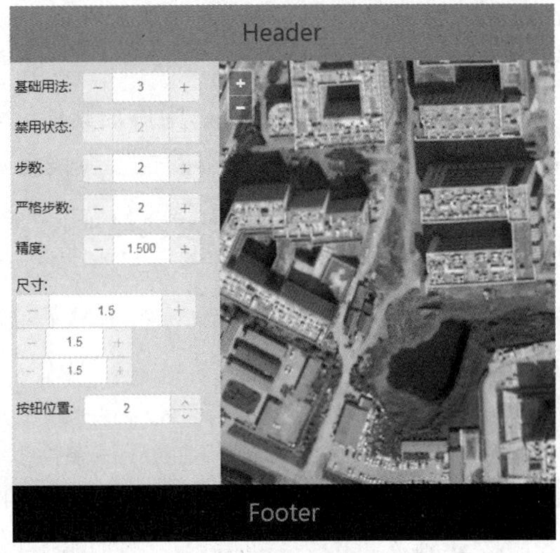

图 3-6　计数器样式

本示例首先创建 Vue 实例，其中 data 中的 num1 至 num6 是各个 el-input-number 标签中 v-model 属性所绑定的变量；methods 中的 handleChange 事件在计数器的数据改变时被触发，其参数接收的是基础用法中计数器内的数据；然后结合 OpenLayers 实现地图的放大缩小功能。代码如下：

```
var vm = new Vue({
    el: "#app",
    data: {
        num1: 3,
        num2: 2,
        num3: 2,
        num4: 2,
        num5: 1.5,
```

```
            num6: 2,
        },
        methods:{
            handleChange:function(val){
                var zoom = map.getView().getZoom();
                map.getView().setZoom(val);
            }
        }
    })
```

（1）基础用法。Element 提供了 el-input-number 标签，使用该标签中的 v-model 属性绑定一个变量即可实现计数器，计数器的取值范围可通过 min 属性和 max 属性来设置。代码如下：

```
<el-input-number v-model="num1" :min="1" :max="20" size="small" @change="handleChange">
```

（2）禁用状态。在 el-input-number 标签中添加 disabled 属性可以使该标签处于禁用状态。代码如下：

```
<el-input-number v-model="num2" disabled size="small">
```

（3）步数。在 el-input-number 标签中设置 step 属性可控制步数。代码如下：

```
<el-input-number v-model="num3" :step="3" size="small">
```

（4）严格步数。严格步数是指在计数器中只能输入步数的倍数，而步数却没有这个要求。在 el-input-number 标签中设置 step-strictly 属性即可实现严格步数。代码如下：

```
<el-input-number v-model="num4" :step="2" step-strictly size="small">
```

（5）精度。如果步长小于 1，则会涉及精度问题。在 el-input-number 标签中设置 precision 属性可以控制数值的精度，precision 属性的值必须是非负整数，并且不能小于 step 的小数位数。代码如下：

```
<el-input-number v-model="num5" :precision="3" :step="0.01" :min="1" :max="2" size="small">
```

（6）尺寸。Element 为计数器提供了 4 种尺寸，分别是 large、medium、small、min，默认的尺寸为 large。在 el-input-number 标签中设置 size 属性可以控制计数器的尺寸。代码如下：

```
<el-input-number v-model="num5" :min="1" :max="2" size="medium"> </el-input-number>
<el-input-number v-model="num5" :min="1" :max="2" size="small"> </el-input-number>
<el-input-number v-model="num5" :min="1" :max="2" size="mini"> </el-input-number>
```

（7）按钮位置。在计数器中，通过设置 el-input-number 标签中的 controls-position 属性可以控制按钮位置。代码如下：

```
<el-input-number v-model="num6" controls-position="right" size="small">
```

3.5 选择器（Select）

Element 为选择器提供了不同的样式与用法，本节将从基础用法、禁用选项、禁用状态、可清空单选、基础多选、自定义模板、分组、可搜索、创建条目等方面讲解选择器的样式与用法。本节首先结合地图展示选择器的样式，如图 3-7 所示；然后对选择器的样式与用法进行讲解。

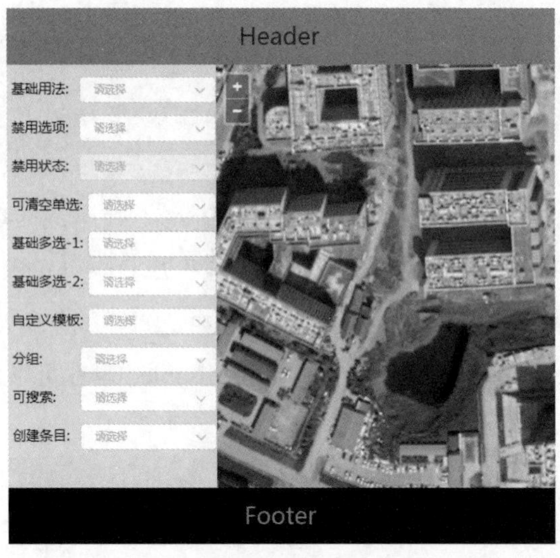

图 3-7　选择器样式

本示例首先创建 Vue 实例，其中 data 中的 value1 至 value9 是各个 el-select 标签中 v-model 属性所绑定的变量，options1 和 options2 是 el-option 标签中进行列表渲染的数据选项，其中 options2 是用于分组样式的选择器，其余的选择器使用 options1 所对应的数据，methods 中的 handleChange 事件在选择器的选中值改变时被触发，回调参数为目前选中的值，该值与 el-option 标签中的 value 属性所对应的值联动；然后结合 OpenLayers 实现地图视角的改变。代码如下：

```
var vm = new Vue({
    el: "#app",
    data: {
        value1: "",
        value2: "",
        value3: "",
        value4: "",
        value5_1: [],
        value5_2: [],
        value6: "",
        value7: "",
        value8: "",
        value9: [],
        options1: [{ val: "一号教学楼", ID: "1" },
                  { val: "二号教学楼", ID: "2", disabled: true },
                  { val: "三号教学楼", ID: "3" }],
        options2: [
```

```
                {
                    label: '食堂',
                    options: [{ val: '一号食堂' }, { val: '二号食堂' }]
                },
                {
                    label: '教学楼',
                    options: [{ val: '一号教学楼' }, { val: '二号教学楼' }]
                },
            ]
        },
        methods:{
            handleChange:function(val){
                if(val==="一号教学楼"){
                    map.getView().setCenter([2700, 1600]);
                }
                else if(val==="二号教学楼"){
                    map.getView().setCenter([3000, 2000]);
                }
                else{
                    map.getView().setCenter([3700, 1000]);
                }
            }
        }
    })
```

（1）基础用法。Element 是通过 el-select 标签和 el-option 标签来实现选择器的，其中 el-select 标签中 v-model 属性所绑定的变量与 el-option 标签中 value 属性的值进行联动，el-option 标签中的 label 属性用于数据显示，若没有 label 属性，则显示 value 属性所对应的值。代码如下：

```
<el-select v-model="value1" placeholder="请选择" size="mini">
<el-option v-for="item of options1" :label="item.val" :value="item.val"></el-option>
</el-select>
```

（2）禁用选项。在 el-option 标签中将 disabled 属性设置为 true，即可实现禁用选项。代码如下：

```
<el-select v-model="value2" placeholder="请选择" size="mini">
    <el-option v-for="item of options1" :label="item.val" :value="item.val"
        :disabled="item.disabled">
    </el-option>
</el-select>
```

（3）禁用状态。在 el-select 标签中设置 disabled 属性，可使选择器处于禁用状态。代码如下：

```
<el-select v-model="value3" placeholder="请选择" size="mini" disabled>
    <el-option v-for="item of options1" :label="item.val" :value="item.val">
    </el-option>
</el-select>
```

（4）可清空单选。在 el-select 标签中设置 clearable 属性，即可使选择器具有可清空单选的功能。需要注意的是，clearable 属性仅适用于单选。代码如下：

```
<el-select v-model="value4" placeholder="请选择" size="mini" clearable>
    <el-option v-for="item of options1" :label="item.val" :value="item.val">
    </el-option>
</el-select>
```

（5）基础多选。在 el-select 标签中设置 multiple 属性可以启动多选功能，此时 v-model 的值是由当前选中值所组成的数组。在默认情况下，选中值会以标签的形式展现，通过设置 collapse-tags 属性可以将选中值合并为一段文字。代码如下：

```
<el-select v-model="value5_1" placeholder="请选择" size="mini" multiple>
    <el-option v-for="item of options1" :label="item.val" :value="item.val">
    </el-option>
</el-select>
```

（6）自定义模板。将自定义的 HTML 模板插入 el-option 中的插槽可以实现自定义模板。代码如下：

```
<el-select v-model="value6" placeholder="请选择" size="mini">
    <el-option v-for="item of options1" :label="item.val" :value="item.val">
        <span style="float: right;">{{ item.ID }}</span>
        <span
        style="float: left; color: #8492a6; font-size: 13px">{{ item.val }}</span>
    </el-option>
</el-select>
```

（7）分组。使用 el-option-group 标签可以对备选项进行分组，该标签中的 label 属性为分组名。代码如下：

```
<el-select v-model="value7" placeholder="请选择" size="mini">
    <el-option-group v-for="group in options2" :label="group.label">
        <el-option v-for="item in group.options" :value="item.val">
        </el-option>
    </el-option-group>
</el-select>
```

（8）可搜索。可搜索是指可以在选择器中输入信息来搜索符合条件的选项，在 el-select 标签中设置 filterable 属性可以启用搜索功能。代码如下：

```
<el-select v-model="value8" filterable placeholder="请选择" size="mini">
    <el-option v-for="item of options1" :value="item.val">
    </el-option>
</el-select>
```

（9）创建条目。在 el-select 标签中使用 allow-create 属性后，可以通过在输入框中输入文字来创建条目，注意此时 filterable 属性必须为 true。代码如下：

```
<el-select v-model="value9" allow-create filterable multiple placeholder="请选择" size="mini">
```

```
        <el-option v-for="item of options1" :value="item.val">
        </el-option>
    </el-select>
```

3.6 级联选择器（Cascader）

Element 为级联选择器提供了不同的样式与用法，本节将从基础用法、禁用选项、可清空、仅显示最后一级、多选、选择任意一级选项、可搜索、自定义节点内容等方面讲解级联选择器的样式与用法。本节首先结合地图展示级联选择器的样式，如图 3-8 所示；然后对级联选择器的样式与用法进行讲解。

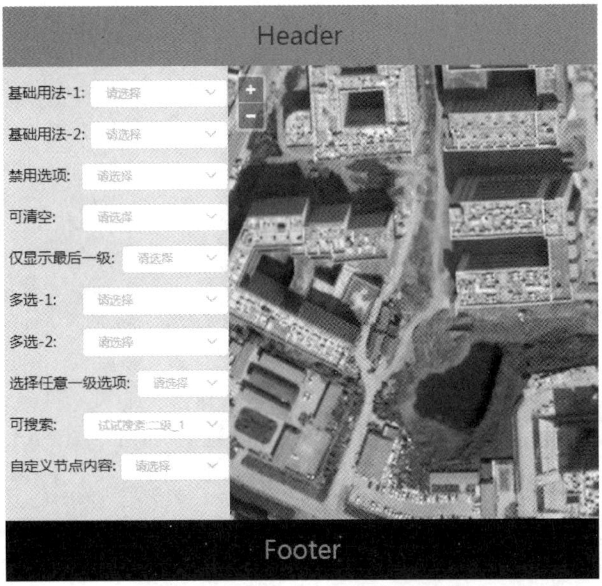

图 3-8 级联选择器样式

本示例首先创建 Vue 实例，其中 data 中的 value1 至 value7 是各个 el-cascader 标签中 v-model 属性所绑定的变量，options1 属性和 options2 属性指定选项数组后即可渲染出一个级联选择器，其中 options2 属性用于禁用选项样式的级联选择器，其余的级联选择器使用 options1 属性对应的数据，methods 中的 handleChange 事件在级联选择器的选中值改变时被触发，回调参数为“基础用法-1”中被选中的值；然后结合 OpenLayers 实现地图视角的改变。代码如下：

```
var vm = new Vue({
    el: "#app",
    data: {
        value1_1: [],
        value1_2: [],
        value2: [],
        value3: [],
```

```
        value4: [],
        value5_1: [],
        value5_2: [],
        value6: [],
        value7: [],
        options1: [{
            value: "公共设施",
            label: "公共设施",
            children: [{
                label: "教学楼",
                value: "教学楼",
                children: [{
                    label: "一号教学楼",
                    value: "一号教学楼",
                },
                {
                    label: "二号教学楼",
                    value: "二号教学楼",
                },
                ]
            },
            {
                label: "学院楼",
                value: "学院楼",
                children: [{
                    label: "地信学院",
                    value: "地信学院",
                },
                {
                    label: "环境学院",
                    value: "环境学院",
                },
                ]
            }]
        }],
        options2: [{
            value: "公共设施",
            label: "公共设施",
            children: [{
                label: "教学楼",
                value: "教学楼",
                children: [{
                    label: "一号教学楼",
                    value: "一号教学楼",
                },
                {
                    label: "二号教学楼",
```

```
                    value: "二号教学楼",
                },
                ]
            },
            {
                label: "学院楼",
                value: "学院楼",
                children: [{
                    label: "地信学院",
                    value: "地信学院",
                },
                {
                    label: "环境学院",
                    value: "环境学院",
                },
                ]
            }]
        }]
    },
    methods: {
        handleChange: function (val) {
            if (val === "地信学院") {
                map.getView().setCenter([2700, 1600]);
            }
            else if (val === "环境学院") {
                map.getView().setCenter([2000, 1600]);
            }
            else if (val === "一号教学楼") {
                map.getView().setCenter([2500, 1600]);
            }
            else {
                map.getView().setCenter([1700, 1600]);
            }
        }
    }
})
```

（1）基础用法。Element 通过 el-cascader 标签提供了一个级联选择器，只需要为 el-cascader 标签中的 options 属性指定选项数组即可渲染出一个级联选择器。Element 为 v-model 绑定了一个变量，通过 props.expandTrigger 定义展开子菜单的触发方式。若 expandTrigger 的值为 hover，则当鼠标光标悬浮到菜单上时便可展开其子菜单；若 expandTrigger 的值为 click，则当鼠标单击菜单时便可展开其子菜单，默认方式是单击触发子菜单。代码如下：

```
<el-cascader v-model="value1_1" :options="options1" size="mini" @change="handleChange"> </el-cascader>
```

（2）禁用选项。在 options 属性指定的数组中，为某个元素设置 disabled 属性并赋值为 true，便可以把该选项设置为禁用选项。代码如下：

```
<el-cascader v-model="value2" :options="options2" size="mini"></el-cascader>
```

（3）可清空。在 el-cascader 标签中设置 clearable 属性可清空输入框的内容。代码如下：

```
<el-cascader v-model="value3" :options="options1" size="mini" clearable></el-cascader>
```

（4）仅显示最后一级。上面的输入框显示的是所有级别元素的完整路径，若只需要输入框显示最后一级元素，则可以在 el-cascader 标签中将 show-all-levels 属性设置为 false。代码如下：

```
<el-cascader v-model="value4" :options="options1" size="mini"
show-all-levels="false"></el-cascader>
```

（5）多选。在 el-cascader 标签中设置 props 属性，将该属性对应的变量（该变量是一个对象）设置为 multiple:true 即可实现多选，在 el-cascader 标签中设置 collapse-tags 属性可以折叠显示选择的元素。代码如下：

```
<el-cascader :options="options1" size="mini" :props="{ multiple: true }" clearable v-model="value5_1" >
</el-cascader>
<el-cascader :options="options1" size="mini" :props="{ multiple: true }" collapse-tags clearable v-model=
"value5_2"></el-cascader>
```

（6）选择任意一级选项。上面所讲到的级联选择器都只能选择子选项，若想要选择任意级别的选项，则可在 el-cascader 标签中设置 props.checkStrictly 属性并赋值为 true。代码如下：

```
<el-cascader v-model="value6" :options="options1" size="mini"
:props="{checkStrictly: true}" clearable></el-cascader>
```

（7）可搜索。在 el-cascader 标签中设置 props 属性，将该属性对应的变量（该变量是一个对象）设置为 checkStrictly:true 即可进行搜索。代码如下：

```
<el-cascader v-model="value7" :options="options1" size="mini" placeholder="试试搜索:二级_1"filterable>
</el-cascader>
```

（8）自定义节点内容。在 el-cascader 标签中添加 template 标签，并在 template 标签中设置 slot-scope 属性可以自定义节点内容，scoped slot 会传入 node 和 data 两个字段，这两个字段分别表示当前节点的 Node 对象和数据。代码如下：

```
<el-cascader :options="options1" size="mini">
    <template slot-scope="{ node, data }">
        <span>{{ data.label }}</span>
        <span v-if="!node.isLeaf"> ({{ data.children.length }}) </span>
    </template>
</el-cascader>
```

3.7 开关（Switch）

Element 为开关提供了不同的样式与用法，本节将从基础用法、文字描述、扩展的 value

类型、禁用状态等方面讲解开关的样式与用法。本节首先结合地图展示开关的样式，如图 3-9 所示；然后对开关的样式与用法进行讲解。

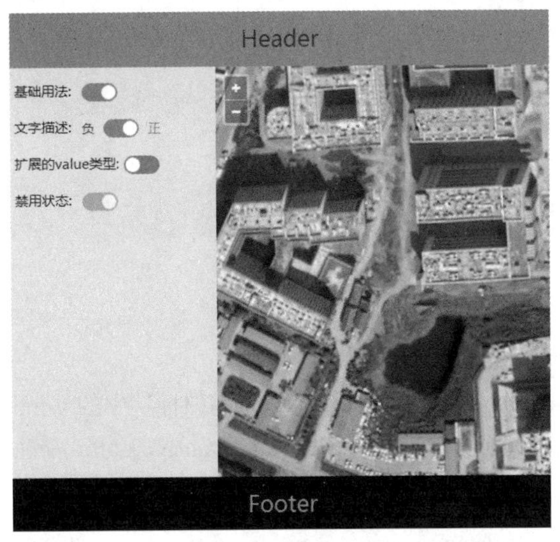

图 3-9　开关样式

本示例首先创建 Vue 实例，其中 data 中的 value1 至 value4 是各个 el-switch 标签中 v-model 属性绑定的变量，methods 中的 handleChange 事件在开关的状态改变时被触发，回调参数为基础用法中开关的当前状态，若开关的状态为开，则返回 true，若开关的状态为关，则返回 false；然后结合 OpenLayers 实现地图的隐藏和显示，单击“基础用法”中的开关可隐藏地图（见图 3-10）或显示地图。代码如下：

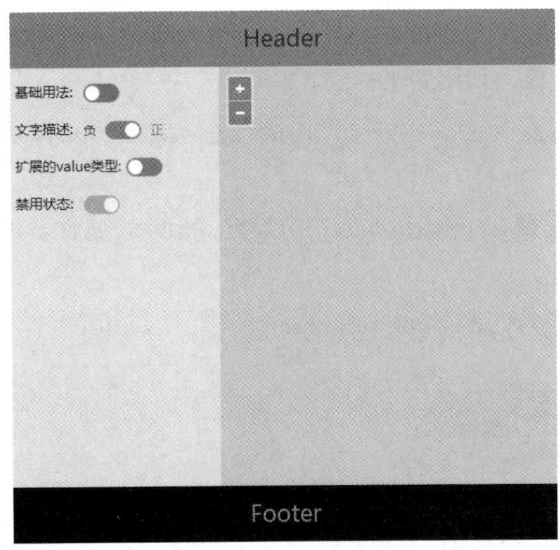

图 3-10　地图隐藏

```
var vm = new Vue({
    el: "#app",
```

```
        data: {
            value1: true,
            value2: true,
            value3: 10,
            value4:true
        },
        methods:{
            handleChange:function(val){
                //imageLayer 图层的显示与否根据开关的值来决定
                imageLayer.setVisible(val);
            }
        }
    })
```

（1）基础用法。Element 提供了 el-switch 标签，为该标签中的 v-model 属性绑定一个 Boolean 类型的变量即可实现开关，通过 active-color 属性与 inactive-color 属性可以设置开关的背景色。代码如下：

```
<el-switch v-model="value1" active-color="#13ce66" inactive-color="#ff4949" @change="handleChange">
```

（2）文字描述。在使用开关时，通过文字可以让用户更加清楚地明白开关所表达的意思，只需要在 el-switch 标签中设置 active-text 属性和 inactive-text 属性并赋予相应的文字即可为开关添加文字描述。代码如下：

```
<el-switch v-model="value2" active-color="#13ce66" inactive-color="#ff4949"  active-text="正" inactive-text="负"></el-switch>
```

（3）扩展的 value 类型。在实现开关时，Element 为 el-switch 标签中的 v-model 属性绑定的是 Boolean 类型的变量，用户可以在 el-switch 标签中设置 active-value 属性和 inactive-value 属性来实现所需的 value 值。代码如下：

```
<el-switch v-model="value3" active-color="#13ce66" inactive-color="#ff4949" active-value="10" inactive-value="0"></el-switch>
```

（4）禁用状态。只需要在 el-switch 标签中设置 disabled 属性即可使开关处于禁用状态。代码如下：

```
<el-switch v-model="value4"  disabled></el-switch>
```

3.8 滑块（Slider）

Element 为滑块提供了不同的样式与用法，本节将从基础用法、离散值、带有输入框、范围选择、竖向模式、展示标注等方面讲解滑块的样式与用法。本节首先结合地图展示滑块的样式，如图 3-11 和图 3-12 所示；然后对滑块的样式与用法进行讲解。

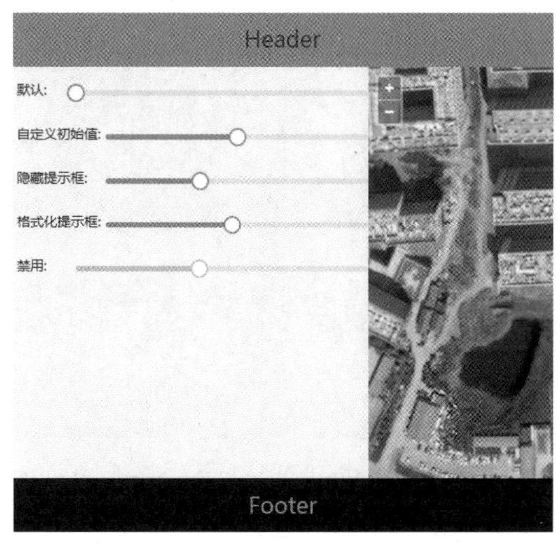

图 3-11　滑块样式（一）

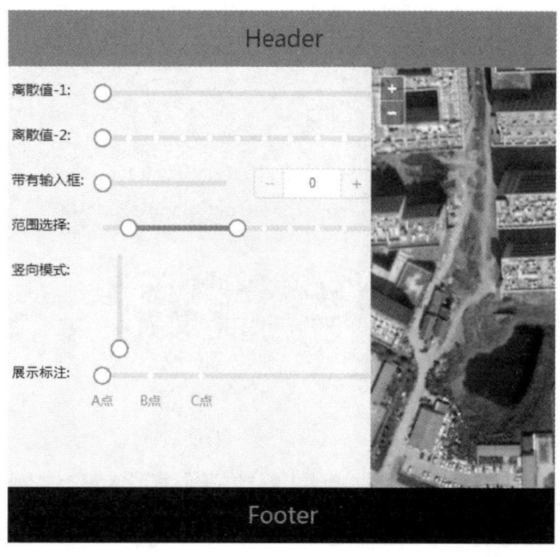

图 3-12　滑块样式（二）

第一个页面（见图 3-11）首先创建了 Vue 实例，其中 data 中的 value1 至 value5 是各个 el-slider 标签中 v-model 属性所绑定的变量，methods 中的 formatTooltip 事件用于格式化提示框的数值，handleChange 事件在滑块的值改变后被触发，回调参数为默认样式的滑块中 v-model 属性所对应的变量值 value1；然后结合 OpenLayers 实现地图透明度的变化（见图 3-13）。代码如下：

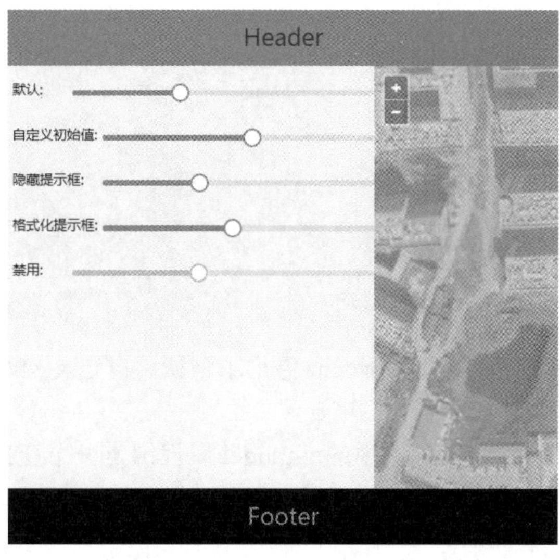

图 3-13　地图透明度的变化

```
var vm = new Vue({
    el: "#app",
    data: {
        value1_1: 50,
        value1_2: 5,
        value1_3: 36,
```

```
            value1_4: 48,
            value1_5: 42
        },
        methods:{
            handleChange:function(val){
                imageLayer.setOpacity(val/100);
            },
            formatTooltip:function(val){
                return val / 100;
            }
        }
    })
```

第二个页面（见图 3-12）创建了 Vue 实例，其中 data 中的 value1 至 value6 是各个 el-slider 标签中 v-model 属性所绑定的变量，methods 中的 formatTooltip 事件用于格式化提示框的数值。代码如下：

```
var vm = new Vue({
    el: "#app",
    data: {
        value2: 0,
        value3: 0,
        value4: 0,
        value5: [10, 50],
        value6: 0,
        value7: 0
    },
    methods: {
        formatTooltip: function (val) {
            return val / 100;
        }
    }
})
```

（1）基础用法。在基础用法中，Element 提供了默认、自定义初始值、隐藏提示框、格式化提示框、禁用等几种样式。

① 默认。Element 为 el-slider 标签中的 v-model 属性绑定一个变量，该变量的值为滑块在滑条上所对应的值。代码如下：

```
<el-slider v-model="value1_1" @change="handleChange"></el-slider>
```

② 自定义初始值。自定义初始值是指将 v-model 属性所对应的变量设置为自定义值。滑条的最大值默认为 100，最小值默认为 0，在 el-slider 标签中设置 max 属性和 min 属性可以自定义滑条的最大值和最小值。代码如下：

```
<el-slider v-model="value1_2" :min="-50" :max="50"></el-slider>
```

③ 隐藏提示框。在默认情况下，当鼠标光标移动到滑块上时，会弹出一个提示框来提示

当前滑块所对应的值。在 el-slider 标签中设置 show-tooltip 属性并赋值为 false 即可隐藏该提示框。代码如下：

```
<el-slider v-model="value1_3" :show-tooltip="false"></el-slider>
```

④ 格式化提示框。当需要将滑块所对应的值在提示框中进行格式化显示时，只需要在 el-slider 标签中设置 format-tooltip 属性即可。该属性所对应的变量为一个方法，该方法会将滑块当前所对应的值作为一个参数，对该参数进行处理可达到格式化提示框的目的。代码如下：

```
<el-slider v-model="value1_4" :format-tooltip="formatTooltip"></el-slider>
```

⑤ 禁用。只需要在 el-slider 标签中设置 disabled 属性即可禁用滑块。代码如下：

```
<el-slider v-model="value1_5" disabled></el-slider>
```

（2）离散值。在使用滑块时，通过改变步长可以间断地显示滑块所对应的值，只需要在 el-slider 标签中设置 step 属性并赋值步长即可。当然，也可以设置 show-stops 属性来使滑条显示间断点。代码如下：

```
<el-slider v-model="value2" :step="10"></el-slider>
<el-slider v-model="value3" :step="10" show-stops></el-slider>
```

（3）带有输入框。带有输入框的滑块可以快速定位到精确数值，只需要在 el-slider 标签中设置 show-input 属性即可实现带有输入框的滑块。代码如下：

```
<el-slider v-model="value4" show-input></el-slider>
```

（4）范围选择。当需要在滑条上取两个节点来选择某一数值范围时，只需要在 el-slider 标签中设置 range 属性，并将 v-model 属性所绑定的变量设置为某一数值范围的数组即可。代码如下：

```
<el-slider v-model="value5" :step="10" range show-stops></el-slider>
```

（5）竖向模式。Element 不仅提供了横向的滑块样式，也提供了竖向的滑块样式。只需要在 el-slider 标签中设置 vertical 属性，并通过 height 属性设置高度即可实现竖向的滑块样式。代码如下：

```
<el-slider v-model="value6" vertical height="100px"></el-slider>
```

（6）展示标注。当需要对滑条的间断点进行标注时，只需要在 el-slider 标签中设置 marks 属性并赋予一个变量即可，该变量对应的是一个对象，在该对象中可对标注的样式进行设置。代码如下：

```
<el-slider v-model="value7" :marks="{0: 'A 点',18: 'B 点',37: 'C 点'}"></el-slider>
```

3.9 时间选择器（TimePicker）

Element 为时间选择器提供了不同的样式与用法，本节将从固定时间点、任意时间点、固定时间范围、任意时间范围等方面讲解时间选择器的样式与用法。本节首先结合地图展示时间选择器的样式，如图 3-14 所示；然后对时间选择器的样式与用法进行讲解。

图 3-14　时间选择器样式

本示例首先创建 Vue 实例，其中 data 中的 value1 至 value4、startTime 和 endTime 是各个标签中 v-model 属性所绑定的变量，vectorSource 是一个文本标注的全局变量，用于显示地图上的文本标注，methods 中的 handleChange 事件在时间选择器的数据改变时被触发，回调参数为固定时间点样式的时间选择器中的数据；然后结合 OpenLayers 实现地图的文本标注功能（见图 3-15）。代码如下：

图 3-15　地图文本标注（一）

```
var vm = new Vue({
    el: "#app",
    data: {
        value1: "",
        value2_1: new Date(2019, 10, 1, 10, 10),
        value2_2: new Date(2019, 10, 1, 10, 10),
        startTime: "",
        endTime: "",
        value3: [new Date(2019, 10, 1, 8, 00), new Date(2019, 10, 1, 9, 00)],
        value4: [new Date(2019, 10, 1, 8, 00), new Date(2019, 10, 1, 9, 00)],
        //矢量标注的数据源
        vectorSource: new ol.source.Vector()
    },
    methods: {
        handleChange: function (val) {
            //创建矢量标注样式函数
            var createLabelStyle = function (feature) {
                return new ol.style.Style({
                    text: new ol.style.Text({
                        //位置
                        textAlign: 'center',
                        //基准线
                        textBaseline: 'middle',
                        //文字样式
                        font: 'normal 18px  微软雅黑',
                        //文本内容
                        text: feature.get('name'),
                        //文本填充样式（即文字颜色）
                        fill: new ol.style.Fill({ color: '#aa4400' }),
                        stroke: new ol.style.Stroke({ color: '#ffcc33', width: 2 })
                    })
                });
            }
            //创建点的坐标
            var pt = [2400, 1900];
            //实例化 Vector 要素，通过矢量图层将实例化的 Vector 要素添加到地图容器中
            var iconFeature = new ol.Feature({
                geometry: new ol.geom.Point(pt),
                name: val,
            })
            iconFeature.setStyle(createLabelStyle(iconFeature));
            //在添加文本标注前先清空原来的文本标注
            this.vectorSource.clear();
            this.vectorSource.addFeature(iconFeature);
            //矢量标注图层
            var vectorLayer = new ol.layer.Vector({
                source: this.vectorSource
```

```
                });
                map.addLayer(vectorLayer);
            }
        }
    })
```

（1）固定时间点。在 Element 提供的 el-time-select 标签中设置 picker-options 属性并赋值一个对象即可实现固定时间选择器。在该对象中，可通过设置 start、step、end 来指定起始时间、步长和结束时间。代码如下：

```
<el-time-select placeholder="选择时间" v-model="value1" :picker-options="{
    start: '06:30',
    step: '00:30',
    end: '17:00'
}" @change="handleChange">
</el-time-select>
```

（2）任意时间点。在 Element 提供的 el-time-picker 标签中设置 picker-options 属性并赋值一个对象，若该对象为空，则可以选择任意时间，若赋值的是该对象的 selectableRange（一个时间范围），则只能在该时间范围内选择时间。v-model 属性所对应变量的值可以通过 new Date 来实例化一个时间。在选择时间时，Element 提供了两种交互方式，在默认情况下是通过鼠标滚轮来选择时间的，打开 arrow-control 属性则通过页面上的箭头来选择时间。代码如下：

```
<!-- 鼠标滚轮选择 -->
<el-time-picker v-model="value2_1" :picker-options="{selectableRange: '06:00:00 - 21:00:00'}" placeholder=
"任意时间点"></el-time-picker>
<!-- 鼠标单击选择 -->
<el-time-picker arrow-control v-model="value2_2" :picker-options="{selectableRange: '06:00:00 - 21:00:00'}"
placeholder="任意时间点">
```

（3）固定时间范围。在固定时间点的基础上，通过两个 el-time-select 标签的组合可以选择一个固定时间范围。需要注意的是，在结束时间的 picker-options 对象中需要设置 minTime 为起始时间的值。代码如下：

```
<el-time-select placeholder="起始时间" v-model="startTime" :picker-options="{
    start: '06:30',
    step: '00:30',
end: '17:00'}"
>
</el-time-select>
<el-time-select placeholder="结束时间" v-model="endTime" :picker-options="{
    start: '06:30',
    step: '00:30',
    end: '17:00',
minTime: startTime}"
>
</el-time-select>
```

（4）任意时间范围。在 el-time-picker 标签中设置 is-range 属性即可选择任意时间范围。和固定时间范围一样，Element 同样支持两种交互方式，即鼠标滚轮选择和鼠标单击选择。代码如下：

```
<el-time-picker is-range v-model="value3" range-separator="至"></el-time-picker>
<el-time-picker is-range arrow-control v-model="value4" range-separator="至"></el-time-picker>
```

3.10　日期选择器（DatePic）

Element 为日期选择器提供了不同的样式与用法，本节将从选择日、选择周、选择月、选择年、多个日期、选择日期范围、选择月份范围等方面讲解日期选择器的样式与用法。本节首先结合地图展示日期选择器的样式，如图 3-16 所示；然后对日期选择器的样式与用法进行讲解。

图 3-16　日期选择器样式

本示例首先创建 Vue 实例，其中 data 中的 value1 至 value7 是各个 el-date-picker 标签中 v-model 属性所绑定的变量，vectorSource 是一个文本标注的全局变量，用于显示地图上的文本标注，methods 中的 handleChange 事件在日期选择器的数据改变时被触发，回调参数为选择日样式的日期选择器中的数据；然后结合 OpenLayers 实现地图文本标注（见图 3-17）。代码如下：

```
var vm = new Vue({
    el: "#app",
    data: {
        value1: "",
        value2: "",
        value3: "",
        value4: "",
```

图 3-17　地图文本标注（二）

```
            value5: "",
            value6: "",
            value7: "",
            //矢量标注的数据源
            vectorSource: new ol.source.Vector()
        },
        methods: {
            handleChange: function (val) {
                //创建矢量标注样式函数
                var createLabelStyle = function (feature) {
                    return new ol.style.Style({
                        text: new ol.style.Text({
                            //位置
                            textAlign: 'center',
                            //基准线
                            textBaseline: 'middle',
                            //文字样式
                            font: 'normal 18px 微软雅黑',
                            //文本内容
                            text: feature.get('name'),
                            //文本填充样式（即文字颜色）
                            fill: new ol.style.Fill({ color: '#aa4400' }),
                            stroke: new ol.style.Stroke({ color: '#ffcc33', width: 2 })
                        })
                    });
                }
                //创建点的坐标
                var pt = [2400, 1900];
                //实例化 Vector 要素，通过矢量图层添加到地图容器中
```

```
                var iconFeature = new ol.Feature({
                    geometry: new ol.geom.Point(pt),
                    name: val,
                })
                iconFeature.setStyle(createLabelStyle(iconFeature));
                //添加文本标注之前先清空原来的标注
                this.vectorSource.clear();
                this.vectorSource.addFeature(iconFeature);
                //矢量标注图层
                var vectorLayer = new ol.layer.Vector({
                    source: this.vectorSource
                });
                map.addLayer(vectorLayer);
            }
        }
    })
```

（1）选择日。Element 通过 el-date-picker 标签对日期选择器进行了封装，其中该标签中的 type 属性要设置为 date。在默认情况下，组件接收并返回 Date 对象，可以使用 format 指定输入框的格式，使用 value-format 指定绑定值的格式。代码如下：

```
    <el-date-picker v-model="value1" type="date" placeholder="选择日期" value-format="yyyy 年 MM 月 dd 日"
@change="handleChange">
```

（2）选择周。选择周的实现方式是：首先将 el-date-picker 标签中的 type 属性赋值为 week，然后在 el-date-picker 标签中设置 format 属性并赋值为 yyyy 年第 WW 周。代码如下：

```
    <el-date-picker v-model="value2" type="week" format="yyyy 年第 WW 周"    placeholder="选择日期">
```

（3）选择月。将 el-date-picker 标签中的 type 属性赋值为 month 即可选择月。代码如下：

```
    <el-date-picker v-model="value3" type="month" placeholder="选择日期"></el-date-picker>
```

（4）选择年。将 el-date-picker 标签中的 type 属性赋值为 year 即可选择年。代码如下：

```
    <el-date-picker v-model="value4" type="year" placeholder="选择日期"></el-date-picker>
```

（5）选择多个日期。将 el-date-picker 标签中的 type 属性赋值为 dates 即可选择多个日期。代码如下：

```
    <el-date-picker v-model="value5" type="dates" placeholder="选择日期"></el-date-picker>
```

（6）选择日期范围。将 el-date-picker 标签中的 type 属性赋值为 daterange 即可选择日期范围，通过 range-separator、start-placeholder 和 end-placeholder 占位符可以使日期选择器更加美观。代码如下：

```
    <el-date-picker  v-model="value6"  type="daterange"  range-separator="至"  start-placeholder="开始日期"
end-placeholder="结束日期"></el-date-picker>
```

（7）选择月份范围。将 el-date-picker 标签中的 type 属性赋值为 monthrange 即可选择月份范围，通过 range-separator、start-placeholder 和 end-placeholder 占位符可以使日期选择器更加

美观。代码如下：

```
<el-date-picker v-model="value7" type="monthrange" range-separator="至" start-placeholder="开始日期" end-placeholder="结束日期"></el-date-picker>
```

3.11 日期时间选择器（DateTimePicker）

Element 为日期时间选择器提供了不同的样式与用法，本节将从日期和时间点、日期和时间范围等方面讲解日期时间选择器的样式与用法。本节首先结合地图展示日期时间选择器的样式，如图 3-18 所示；然后对日期时间选择器的样式与用法进行讲解。

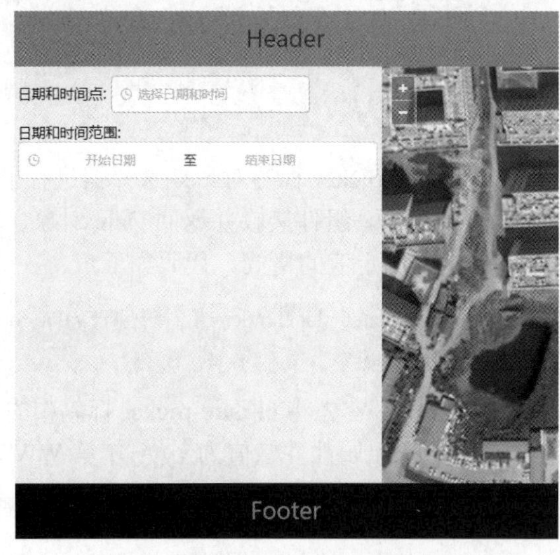

图 3-18　日期时间选择器样式

本示例首先创建 Vue 实例，其中 data 中的 value1 和 value2 是各个 el-date-picker 标签中 v-model 属性所绑定的变量，vectorSource 是一个文本标注的全局变量，用于显示地图上的文本标注，methods 中的 handleChange 事件在日期时间选择器的数据改变时被触发，回调参数为日期和时间点样式的日期时间选择器中的数据；然后结合 OpenLayers 实现地图文本标注（见图 3-19）。代码如下：

```
var vm = new Vue({
    el: "#app",
    data: {
        value1: "",
        value2: "",
        //矢量标注的数据源
        vectorSource: new ol.source.Vector()
    },
    methods: {
        handleChange: function (val) {
```

图 3-19　地图文本标注（三）

```
//创建矢量标注样式函数
var createLabelStyle = function (feature) {
    return new ol.style.Style({
        text: new ol.style.Text({
            //位置
            textAlign: 'center',
            //基准线
            textBaseline: 'middle',
            //文字样式
            font: 'normal 18px  微软雅黑',
            //文本内容
            text: feature.get('name'),
            //文本填充样式（即文字颜色）
            fill: new ol.style.Fill({ color: '#aa4400' }),
            stroke: new ol.style.Stroke({ color: '#ffcc33', width: 2 })
        })
    });
}
//创建点的坐标
var pt = [2450, 1900];
//实例化 Vector 要素，通过矢量图层添加到地图容器中
var iconFeature = new ol.Feature({
    geometry: new ol.geom.Point(pt),
    name: val,
})
iconFeature.setStyle(createLabelStyle(iconFeature));
//添加文本标注之前先清空原来的标注
this.vectorSource.clear();
this.vectorSource.addFeature(iconFeature);
//矢量标注图层
var vectorLayer = new ol.layer.Vector({
    source: this.vectorSource
});
map.addLayer(vectorLayer);
```

```
            }
        }
    })
```

（1）日期和时间点。Element 通过 el-date-picker 标签对日期选择器进行了封装，将该标签中的 type 属性设置为 datetime 即可实现日期时间选择器。在选择时间时将某个时间作为默认时间，将 el-date-picker 标签中的 default-time 属性设置为某一时间也可以实现日期时间选择器。代码如下：

```
    <el-date-picker v-model="value1" type="datetime" placeholder="选择日期和时间" value-format= "yyyy-MM-dd HH:mm:ss"
    @change="handleChange"></el-date-picker>
```

（2）日期和时间范围。将 el-date-picker 标签中的 type 属性赋值为 datetimerange 即可设置日期和时间范围，通过 range-separator、start-placeholder 和 end-placeholder 占位符可以使日期时间选择器更加美观，通过 default-time 属性可以设置默认的起始时间与结束时间。代码如下：

```
    <el-date-picker v-model="value2" type="datetimerange" range-separator="至" start-placeholder="开始日期"
    end-placeholder="结束日期"
    :default-time="['12:00:00', '08:00:00']">
    </el-date-picker>
```

3.12 上传（Upload）

Element 为上传提供了不同的样式与用法，本节将从单击上传、用户头像上传、照片墙、上传文件列表控制、拖曳上传等方面讲解上传的样式与用法。在讲解上传的样式与用法之前，首先要明白该功能需要结合前端与后台，本节通过创建 ASP.NET 中最基础的一般处理程序来完成该功能。下面简单地介绍一下一般处理程序的创建与后台程序的编写。

（1）打开 Visual Studio 2012（版本可自选），选择“文件”→“新建”→“网站”，可打开网站创建页面，如图 3-20 所示。

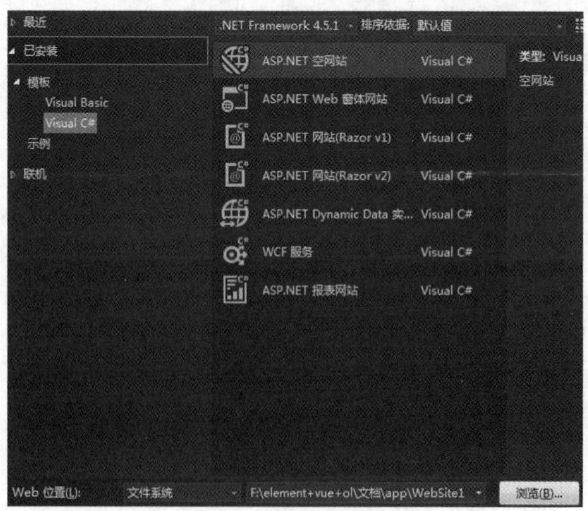

图 3-20　网站创建页面

（2）选择 Visual C#下的 ASP.NET 空网站，单击“浏览”按钮，选择一个目录创建网站，本示例所创建位置与上传的 HTML 页面在同一目录下，如图 3-21 所示。

（3）单击“打开”按钮，可打开“网站已存在”对话框，如图 3-22 所示。

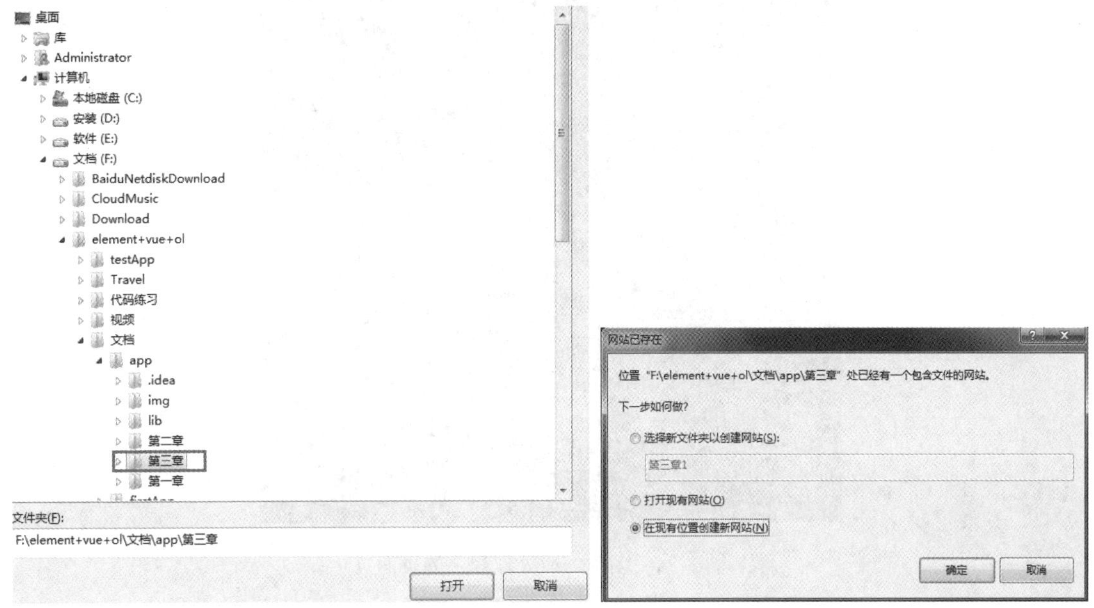

图 3-21　选择创建网站的目录　　　　图 3-22　“网站已存在”对话框

（4）在图 3-22 中选择“在现有位置创建新网站”后单击“确定”按钮，即可创建一个空网站，如图 3-23 所示。

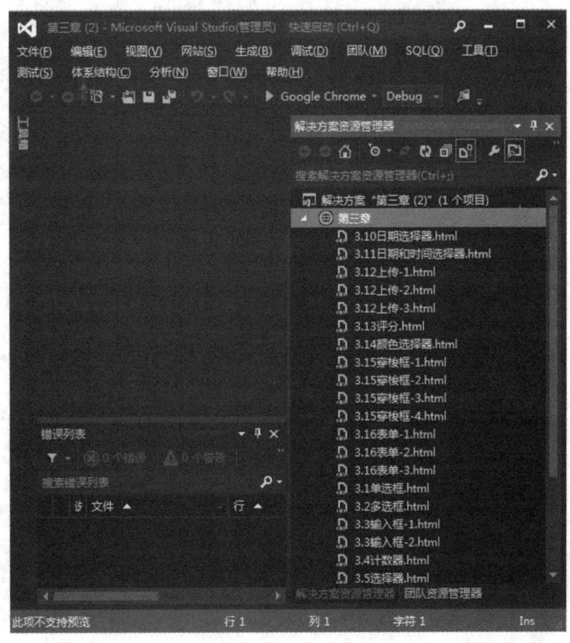

图 3-23　创建一个空网站

（5）右键单击图 3-23 中的“第三章”，在弹出的右键菜单中选择“添加”→“添加新项”，可打开创建一般处理程序的页面，如图 3-24 所示。

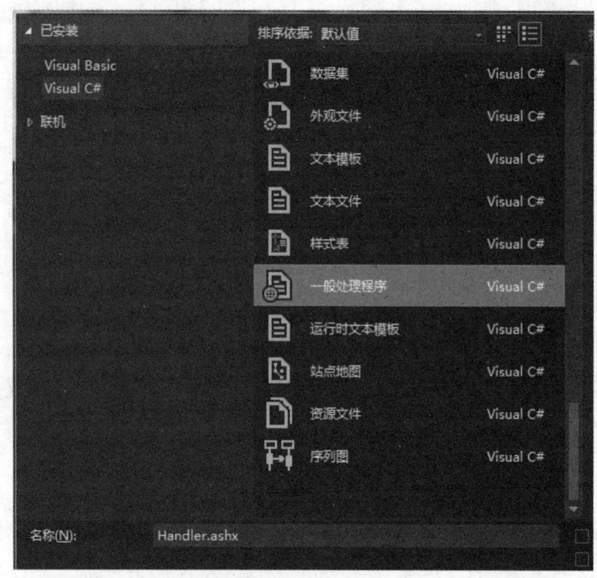

图 3-24　创建一般处理程序的页面

（6）选择“Visual C#”下的“一般处理程序”，单击“添加”按钮，即可成功创建一般处理程序（Handler.ashx），如图 3-25 所示。

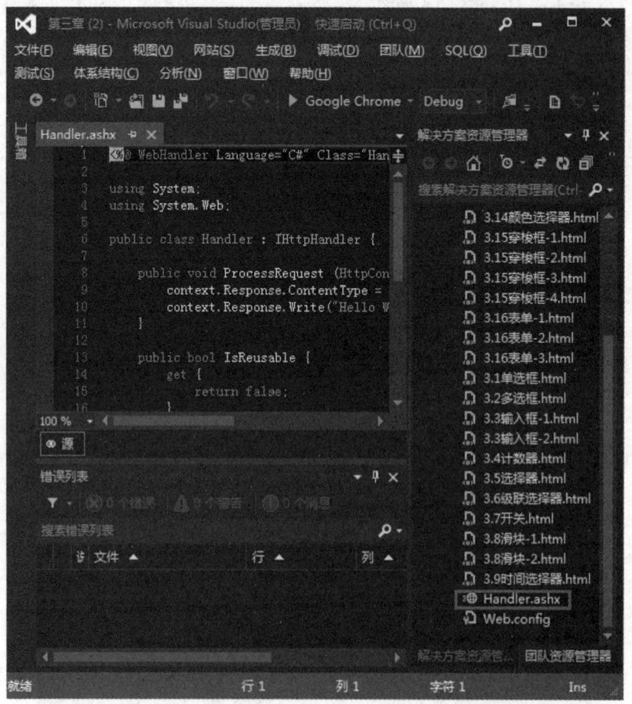

图 3-25　成功创建一般处理程序

（7）将以下代码复制到文件 Handler.ashx 中。

```
<%@ WebHandler Language="C#" Class="Handler" %>
using System;
using System.Web;
public class Handler : IHttpHandler {
    public void ProcessRequest (HttpContext context) {
    HttpFileCollection files = HttpContext.Current.Request.Files;
    for(int i = 0; i < files.Count; i++)
    {
        HttpPostedFile file = files[i];
        //将要上传的文件保存到本地的位置
         file.SaveAs(@"F:\element+vue+ol\文档\app\第三章\files\" + file.FileName);
    }
    }
    public bool IsReusable {
        get {
            return false;
        }
    }
}
```

至此，后台程序编写完毕。接下来对上传的样式与用法进行讲解。由于物理空间的限制，这里以图 3-26、图 3-27 和图 3-28 所示的页面为例来介绍上传的样式与用法。需要注意的是，在使用一般处理程序时，需要在 Visual Studio 2012 中打开 HTML 文件，否则很可能会报跨域错误。

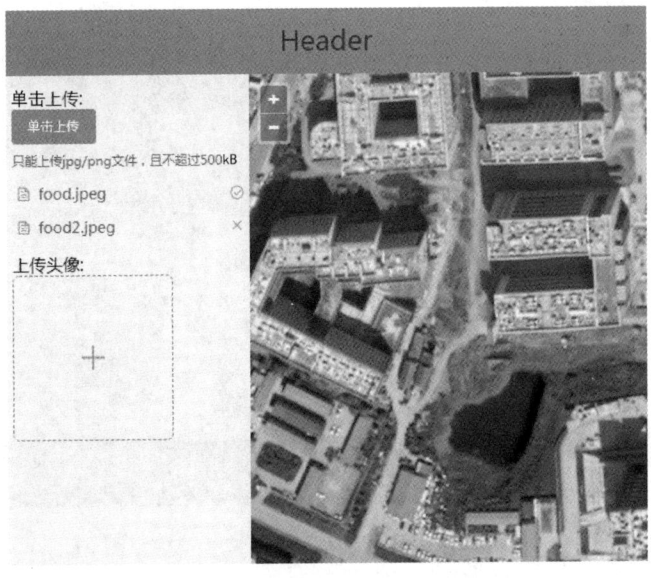

图 3-26　上传样式（一）

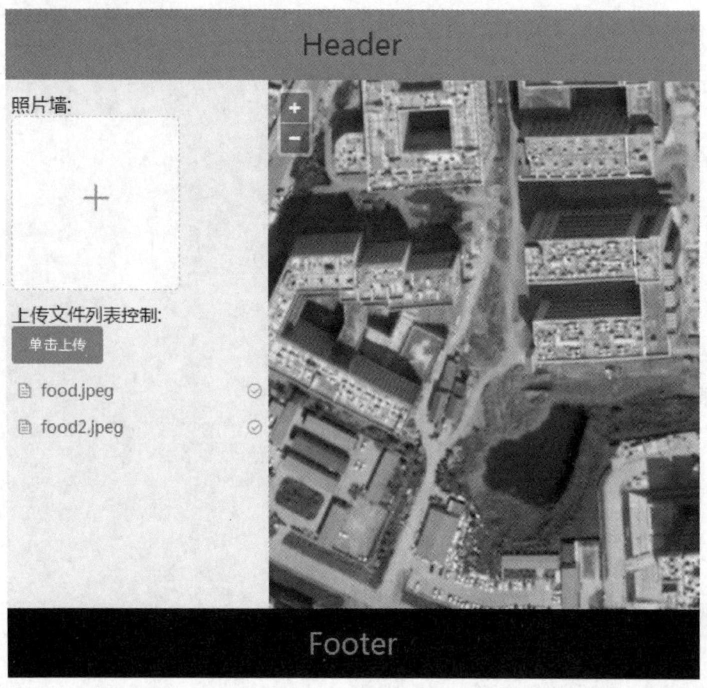

图 3-27　上传样式（二）

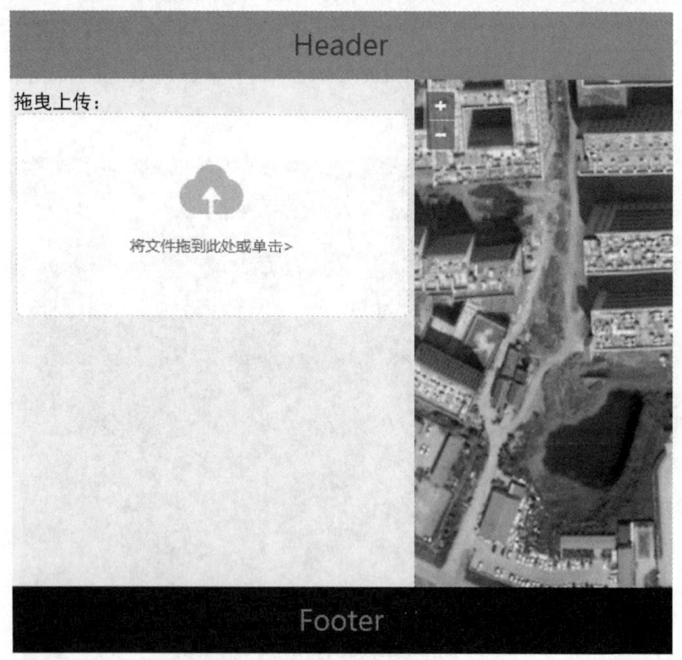

图 3-28　上传样式（三）

对于图 3-26 所示的页面，本示例创建了 Vue 实例，其中 data 中的 num 为上传文件的限制数量，fileList 为初始化页面的文件目录，methods 中的 handleExceed 为文件数量超出限制时的行为，beforeRemove 为文件移除前的响应事件，handleSuccess 为文件上传成功的响应事件，handleError 为文件上传失败的响应事件，handlePicSuccess 为头像上传成功的响应事件，beforePicUpload 为头像上传之前的响应事件。代码如下：

```
var vm = new Vue({
    el: "#app",
    data: {
        num: 3,
        fileList: [{ name: 'test1.txt', url: './files/test1.txt' }, { name: 'test2.txt', url: './files/test2.txt' }],
        imageUrl: '',
    },
    methods: {
        //超出文件数量限制的响应事件
        handleExceed(files, fileList) {
            alert('当前限制选择' + this.num + '个文件，本次选择了 ${files.length} 个文件，共选择了
${files.length + fileList.length} 个文件');
        },
        //文件移除前的响应事件
        beforeRemove(file, fileList) {
            alert('确定移除 ${file.name}？');
        },
        //文件上传成功的响应事件
        handleSuccess(response, file, fileList) {
            alert("上传成功!")
        },
        //文件上传失败的响应事件
        handleError(err, file, fileList) {
            alert("上传失败!")
        },
        //头像上传成功的响应事件
        handlePicSuccess(res, file) {
            this.imageUrl = URL.createObjectURL(file.raw);
        },
        //头像上传之前的响应事件
        beforePicUpload(file) {
            var isJPG = file.type === 'image/png';
            var isSize = file.size / 1024 / 1024 < 3;
            if (!isJPG) {
                alert('上传头像图片只能是 PNG 格式!');
            }
            if (!isSize) {
                alert('上传头像图片大小不能超过 3MB!');
            }
            return isJPG && isSize;
        }
    }
})
```

对于图 3-27 所示的页面，本示例创建了 Vue 实例，其中 data 中的 dialogImageUrl 为照片墙预览时的地址，dialogVisible 用于控制对话框的显示与否，fileList 为初始化页面的文件目录；methods 中的 handleRemove 为照片墙中文件被移除的响应事件，handlePictureCardPreview 为单击文件列表中已上传的文件时的响应事件，handleChange 为文件列表改变时的响应事件。代码如下：

```
var vm = new Vue({
    el: "#app",
    data: {
        dialogImageUrl: '',
        dialogVisible: false,
        fileList: [{ name: 'test1.txt', url: './files/test1.txt' }, { name: 'test2.txt', url: './files/test2.txt' }],
    },
    methods: {
        //照片墙中文件被移除的响应事件
        handleRemove(file, fileList) {
            console.log(file, fileList);
        },
        //单击文件列表中已上传的文件时的响应事件
        handlePictureCardPreview(file) {
            this.dialogImageUrl = file.url;
            this.dialogVisible = true;
        },
        //文件列表改变时的响应事件
        handleChange(file, fileList) {
            console.log(file);
            console.log(fileList);
        }
    }
})
```

对于图 3-28 所示的页面，本示例创建了 Vue 实例，其中 data 中的 fileList 为初始化页面的文件目录，methods 中的 handleSuccess 为文件上传成功的响应事件，handleError 为文件上传失败的响应事件。代码如下：

```
var vm = new Vue({
    el: "#app",
    data: {
        fileList: [{ name: 'test1.txt', url: './files/test1.txt' }, { name: 'test2.txt', url: './files/test2.txt' }],
    },
    methods: {
        //文件上传成功的响应事件
        handleSuccess(response, file, fileList) {
            alert("上传成功!")
        },
        //文件上传失败的响应事件
        handleError(err, file, fileList) {
```

```
            alert("上传失败!")
        },
    }
})
```

（1）单击上传。Element 通过 el-upload 标签对上传进行了封装。在单击上传中，multiple 属性用于决定是否支持多选文件，action 属性对应的是相应的目录，也就是前面编写的一般处理程序的相对目录（也称为相对路径），before-remove 属性对应的是删除文件之前的勾子函数，on-exceed 属性对应的是文件超出个数限制时的勾子函数，limit 属性是最大允许上传的个数，on-success 属性对应的是上传成功时的勾子函数，on-error 属性对应的是上传失败时的勾子函数，其他 API 请参考 Element 官网。代码如下：

```
<el-upload class="upload-demo" action="./Handler.ashx"
    :before-remove="beforeRemove" multiple :limit="num" :on-exceed="handleExceed"
    :on-success="handleSuccess" :on-error="handleError" :file-list="fileList">
    <el-button size="small" type="primary">单击上传</el-button>
    <div slot="tip" class="el-upload__tip">只能上传 jpg/png 文件，且不超过 500kB</div>
</el-upload>
```

（2）用户头像上传。用户头像上传是常见的上传样式之一，用户可以通过单击鼠标来上传自己的头像，el-upload 标签中的 before-upload 属性对应的是上传图片之前的勾子函数，在该勾子函数中可以限制用户上传图片的格式和大小。代码如下：

```
<el-upload class="pic-uploader" action="./Handler.ashx"
    :show-file-list="false" :on-success="handlePicSuccess" :before-upload="beforePicUpload">
    <img v-if="imageUrl" :src="imageUrl" class="pic">
    <i v-else class="el-icon-plus pic-uploader-icon"></i>
</el-upload>
```

（3）照片墙。Element 除了可以上传头像，还可以上传多个图片，即照片墙。list-type 属性可以用来设置文件列表的样式，该属性的取值有 text、picture 和 picture-card，其中 text 表示文件列表样式，picture 表示缩略图列表样式，picture-card 表示图片本身样式。注意，需要在 el-upload 标签下设置一个 el-dialog 标签，el-dialog 标签用于在放大图片时显示该图片的放大版。代码如下：

```
<el-upload action="./Handler.ashx" list-type="picture-card"
    :on-preview="handlePictureCardPreview" :on-remove="handleRemove">
    <i class="el-icon-plus"></i>
</el-upload>
```

（4）上传文件列表控制。在 el-upload 标签中，on-change 属性表示文件状态改变时的勾子函数，在添加文件、上传成功和上传失败时都会调用该勾子函数，根据该勾子函数可对列表进行控制。代码如下：

```
<el-upload action="./Handler.ashx" :on-change="handleChange"
    :file-list="fileList">
    <el-button size="small" type="primary">单击上传</el-button>
</el-upload>
```

（5）拖曳上传。只需要在 el-upload 标签中添加 drag 属性就可以实现拖曳上传，当然拖曳上传并不会与单击上传发生冲突，同样可以使用单击上传。代码如下：

```
<el-upload drag action="./Handler.ashx" multiple :file-list="fileList"
      :on-success="handleSuccess" :on-error="handleError">
      <i class="el-icon-upload"></i>
      <div class="el-upload__text">将文件拖到此处或单击</div>
</el-upload>
```

3.13 评分（Rate）

Element 为评分提供了不同的样式与用法，本节将从默认、区分颜色、辅助文字、只读等方面讲解评分的样式与用法。本节首先结合地图展示评分的样式，如图 3-29 所示；然后对评分的样式与用法进行讲解。

图 3-29　评分样式

本示例首先创建 Vue 实例，其中 data 中的 value1 至 value3 是各个 el-rate 标签中 v-model 属性所绑定的变量，colors 是区分颜色的评分样式中的颜色选择数组，methods 中的 handleChange 事件在评分改变时被触发；然后结合 OpenLayers 实现地图缩放（见图 3-30）。代码如下：

```
var vm = new Vue({
      el: "#app",
      data: {
            value1_1: 3,
            value1_2: null,
            value2: null,
            value3: 3.5,
```

```
            colors: ['#CCC', '#FFCCCC', '#FFCC00'],
        },
        methods:{
            handleChange:function(){
                //地图的缩放级别为当前所选分数
                map.getView().setZoom(this.value1_1);
            }
        }
    })
```

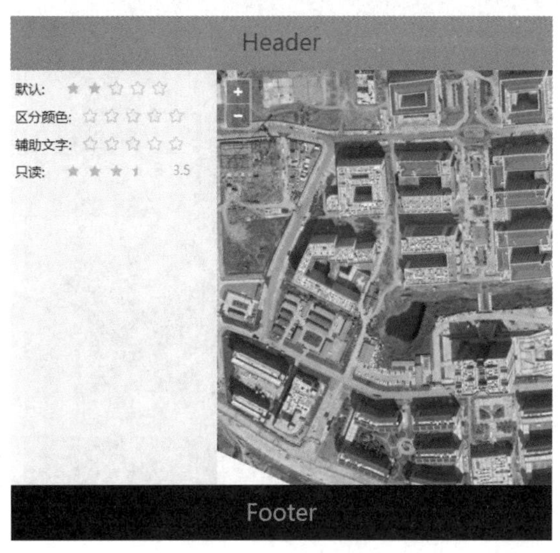

图 3-30　地图缩放

（1）默认和区分颜色。Element 通过 el-rate 标签封装了评分，在 el-rate 标签中通过 v-model 属性绑定了一个用于监听数据的变量。在默认情况下，评分的图标是不区分颜色的。若要区分颜色，则可在 el-rate 标签中设置 colors 属性，从而为不同等级的分数赋予不同的颜色。需要注意的是，若为 colors 属性赋值的是数组，则该数组共有 3 个元素，3 个元素对应 3 个分段的颜色；若为 colors 属性赋值的是对象，则可自定义分段，键为分段的界限值，值为分段对应的颜色。代码如下：

```
<el-rate v-model="value1_1" @change="handleChange"></el-rate>
<el-rate v-model="value1_2" :colors="colors"></el-rate>
```

（2）辅助文字。通过辅助文字可以使评分更加直观，只需要在 el-rate 中设置 show-text 属性即可实现辅助文字。代码如下：

```
<el-rate v-model="value2" show-text></el-rate>
```

（3）只读。为 el-rate 标签设置 disabled 属性即可将评分设置为只读。此时若设置 show-score，则会在右侧显示目前的分值，并且支持小数分值。代码如下：

```
<el-rate v-model="value3" disabled show-score text-color="#ff9900">
```

3.14 颜色选择器（ColorPicker）

Element 为颜色选择器提供了不同的样式与用法，本节将从基础用法、选择透明度、预定义颜色、不同尺寸等方面讲解颜色选择器的样式与用法。本节首先结合地图展示颜色选择器的样式，如图 3-31 所示；然后对颜色选择器的样式与用法进行讲解。

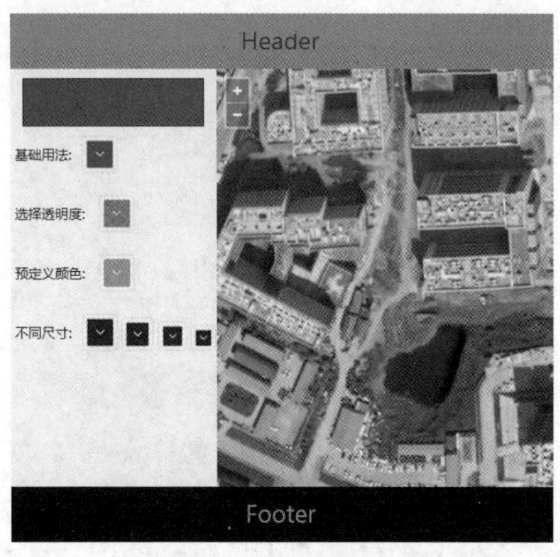

图 3-31　颜色选择器样式

本示例首先创建 Vue 实例，data 中的 styleObj 为顶部矩形框的背景色，color1 至 color3 是各个 el-color-picker 标签中 v-model 属性所绑定的变量，predefineColors 是预定义颜色数组，methods 中的 handleChange 事件在颜色选择器改变时被触发，回调参数为基础用法中颜色选择器的当前值；然后结合 Vue 实现矩形框背景色的改变（见图 3-32）。代码如下：

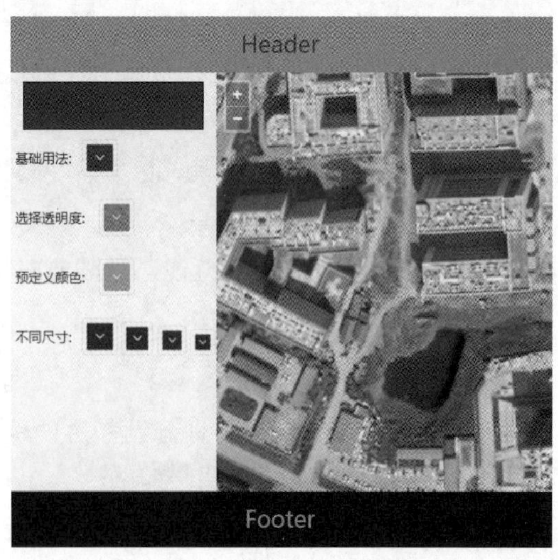

图 3-32　矩形框背景色的改变

```
var vm = new Vue({
    el: "#app",
    data: {
        //该样式用于控制矩形框的背景色
        styleObj:{
            background:"rgb(255,0,0)"
        },
        color1: "rgb(255,0,0)",
        color2: 'rgba(59, 106, 145, 0.8)',
        predefineColors: ["rgb(0,0,0)", "rgb(255,0,0)", "rgb(0,255,0)", "rgb(0,0,255)", "rgba(255,255,0,0.5)"],
        color3:"rgb(0,255,0)",
        color4:"rgb(0,0,255)"
    },
    methods:{
        //通过颜色选择器改变时的响应事件来获取当前选中的颜色并传递给 div 的背景色
        handeChange:function(color){
            this.styleObj.background=color;
        }
    }
})
```

（1）基础用法。Element 通过 el-color-picker 标签提供了颜色选择器，通过 v-model 绑定的变量来监听颜色数据。若绑定的变量对应的值为空，则表示无默认值；否则表示具有一个默认颜色。代码如下：

```
<el-color-picker v-model="color1" @change=handeChange></el-color-picker>
```

（2）选择透明度。在 el-color-picker 标签中添加 show-alpha 属性可以选择透明度。代码如下：

```
<el-color-picker v-model="color2" show-alpha></el-color-picker>
```

（3）预定义颜色。在 el-color-picker 标签中设置 predefine 属性，通过该属性对应的数组可以预定义颜色。代码如下：

```
<el-color-picker v-model="color3" :predefine="predefineColors"></el-color-picker>
```

（4）不同尺寸。通过 size 属性可以定义颜色选择器的尺寸，size 的值可为 large、medium、small 和 mini，默认尺寸为 large。代码如下：

```
<el-color-picker v-model="color4"></el-color-picker>
<el-color-picker v-model="color4" size="medium"></el-color-picker>
<el-color-picker v-model="color4" size="small"></el-color-picker>
<el-color-picker v-model="color4" size="mini"></el-color-picker>
```

3.15 穿梭框（Transfer）

Element 为穿梭框提供了不同的样式与用法，本节将从基础用法、可搜索、可自定义、数据项属性别名等方面讲解穿梭框的样式与用法。由于穿梭框所占物理空间较大，因此本节分别用一个页面布局来对穿梭框的样式与用法进行讲解。

（1）基础用法。Element 通过 el-transfer 标签对穿梭框进行了封装，穿梭框的数据是通过 data 属性传入的。数据是一个对象数组，其中，key 为数据的唯一性标识，label 为显示的文本，disabled 表示该项数据是否禁止转移。在目标列表中，通过 v-model 绑定的变量与 data 中的 key 进行同步对应。left-default-checked 属性和 right-default-checked 属性决定了某些数据项在初始化时是否被勾选。

本示例首先创建 Vue 实例，data 中的 value 是 el-transfer 标签中 v-model 属性所绑定的变量，option 内的对象数组为穿梭框中被渲染的数据；然后结合地图展示穿梭框基础用法的样式（见图 3-33）。代码如下：

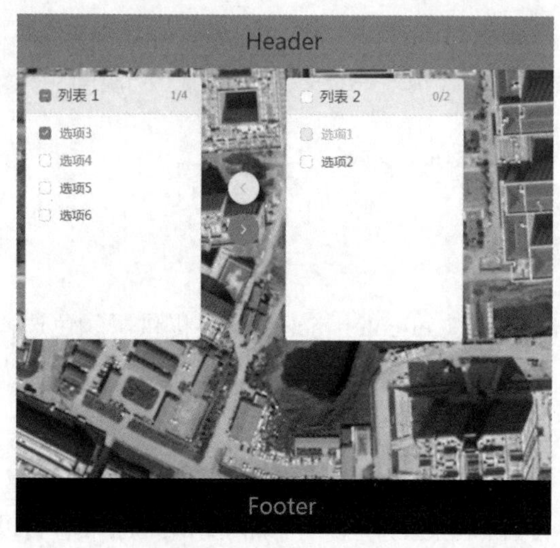

图 3-33　穿梭框基础用法的样式

```
var vm = new Vue({
    el: "#app",
    data: {
        value: [1,2],
        data: [
            { key: 1, label: '选项 1', disabled: true },
            { key: 2, label: '选项 2', disabled: false },
            { key: 3, label: '选项 3', disabled: false },
            { key: 4, label: '选项 4', disabled: false },
            { key: 5, label: '选项 5', disabled: false },
            { key: 6, label: '选项 6', disabled: false }]
    }
})
```

穿梭框基础用法样式对应的标签代码如下：

```
<el-transfer v-model="value" :data="option" :left-default-checked="[2, 3]" :right-default-checked="[1]">
```

（2）可搜索。当穿梭框的数据量非常大时，逐条查询显然就行不通了，可搜索的优势就显现出来了。在 el-transfer 标签中设置 filterable 属性为 true 即可开启搜索功能。在默认情况下，搜索的是穿梭框中 label 属性对应的关键字，也可以使用 filter-method 自定义搜索逻辑，如搜索穿梭框中其他属性对应的关键字。filter-method 属性对应的方法会接收两个参数，第一个参数是当前查询的对象，第二个参数是整个数据项。

本示例首先创建 Vue 实例，data 中的 value 是 el-transfer 标签中 v-model 属性所绑定的变量，option 内的对象数组为穿梭框中被渲染的数据，methods 中的 filterMethod 方法为筛选数据项功能；然后结合地图展示穿梭框可搜索的样式（见图 3-34）。代码如下：

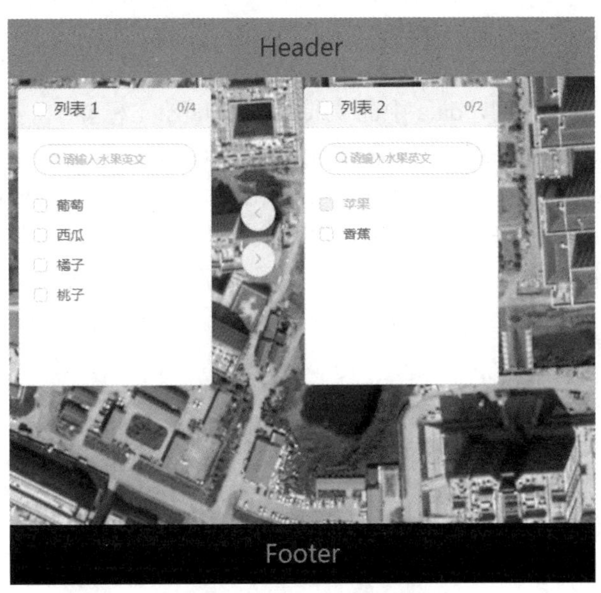

图 3-34　穿梭框可搜索的样式

```
var vm = new Vue({
    el: "#app",
    data: {
        value: [1,2],
        option: [
            { key: 1, label: '苹果', english: "apple", disabled: true },
            { key: 2, label: '香蕉', english: "banana", disabled: false },
            { key: 3, label: '橘子', english: "orange", disabled: false },
            { key: 4, label: '桃子', english: "peach", disabled: false },
            { key: 5, label: '葡萄', english: "garpe", disabled: false },
            { key: 6, label: '西瓜', english: "watermelon", disabled: false }]
    },
    methods: {
        filterMethod: function (queryData, allData) {
            return allData.english.indexOf(queryData) > -1;
        }
    }
})
```

穿梭框可搜索样式对应的标签代码如下：

```
<el-transfer v-model="value" :data="option" :filterable="true"
:filter-method="filterMethod" filter-placeholder="请输入水果英文">
</el-transfer>
```

（3）可自定义。在某些场合下，需要对穿梭框的列表标题文案、按钮文案、数据项的渲染函数、列表底部的勾选状态文案、列表底部的内容区等进行自定义。在 el-transfer 标签中设置 titles、button-texts、render-content 和 format 属性分别对列表标题文案、按钮文案、数据项的渲染函数和列表底部的勾选状态文案进行自定义。对于列表底部的内容区，可通过 left-footer 和 right-footer 属性进行自定义。

本示例首先创建 Vue 实例，data 中的 value 是 el-transfer 标签中 v-model 属性所绑定的变量，option 内的对象数组为穿梭框中被渲染的数据，likeFruits 和 hateFruits 对应的数组与单击事件进行联动，methods 中的 filterMethod 方法为筛选数据项功能，handleChange 在穿梭框数据改变时被触发，handleLikeBtnClick 在单击喜欢的水果按钮时被触发，handleHateBtnClick 在单击讨厌的水果按钮时被触发；然后结合地图展示穿梭框可自定义的样式（见图 3-35）。在该示例中，通过移动水果后，单击喜欢的水果按钮或者讨厌的水果按钮，便可弹出所对应的水果。代码如下：

图 3-35　穿梭框可自定义的样式

```
var vm = new Vue({
    el: "#app",
    data: {
        value: [1, 2],
        option: [
            { key: 1, label: '苹果', english: "apple", disabled: false },
            { key: 2, label: '香蕉', english: "banana", disabled: false },
            { key: 3, label: '橘子', english: "orange", disabled: false },
            { key: 4, label: '桃子', english: "peach", disabled: false },
```

```
            { key: 5, label: '葡萄', english: "garpe", disabled: false },
            { key: 6, label: '西瓜', english: "watermelon", disabled: false }],
        //喜欢的水果
        likeFruits: ['橘子', '桃子', '葡萄', '西瓜'],
        //讨厌的水果
        hateFruits: ['苹果', '香蕉']
    },
    methods: {
        filterMethod: function (queryData, allData) {
            return allData.english.indexOf(queryData) > -1;
        },
        handleLikeBtnClick: function () {
            alert("喜欢的水果是:"+this.likeFruits);
        },
        handleHateBtnClick: function () {
            alert("讨厌的水果是:"+this.hateFruits);
        },
        handleChange: function () {
            //数据改变后，首先清空喜欢的水果和讨厌的水果对应的数组
            this.likeFruits = [];
            this.hateFruits = [];
            //遍历 value 和 data 来获取讨厌的水果
            this.value.forEach((value, index) => {
                this.data.forEach((val, ind) => {
                    if (val.key === value) {
                        this.hateFruits.push(val.label);
                    }
                })
            })
            //将 data 中的所有水果存放到 fruits 数组中
            var fruits = [];
            this.data.forEach((value, index) => {
                fruits.push(value.label);
            })
            //遍历 fruits 数组，去除讨厌的水果，剩下的就是喜欢的水果
            fruits.forEach((value, index) => {
                if (this.hateFruits.indexOf(value) === -1) {
                    this.likeFruits.push(value);
                }
            })
        }
    },
})
```

穿梭框可自定义样式对应的标签代码如下：

```
<el-transfer
```

```
        v-model="value"
        :data="option"
        :filterable="true"
        :filter-method="filterMethod"
        filter-placeholder="请输入水果英文"
        :titles="['爱吃的水果', '不爱吃的水果']"
        :button-texts="['左边去', '右边去']"
        @change="handleChange">
        <el-button
            class="transfer-footer"
            slot="left-footer"
            size="small"
            @click="handleLikeBtnClick">喜欢的水果
        </el-button>
        <el-button
            class="transfer-footer"
            slot="right-footer"
            size="small"
            @click="handleHateBtnClick">
            讨厌的水果
        </el-button>
    </el-transfer>
```

（4）数据项属性别名。在默认情况下，Transfer 仅能识别数据项中的 key、label 和 disabled 字段。如果数据的字段名不同，则可以使用 props 属性为它们设置别名。

本示例首先创建 Vue 实例，data 中的 value 是 el-transfer 标签中 v-model 属性所绑定的变量，option 内的对象数组为穿梭框中被渲染的数据；然后结合地图展示穿梭框数据项属性别名的样式（见图 3-36）。代码如下：

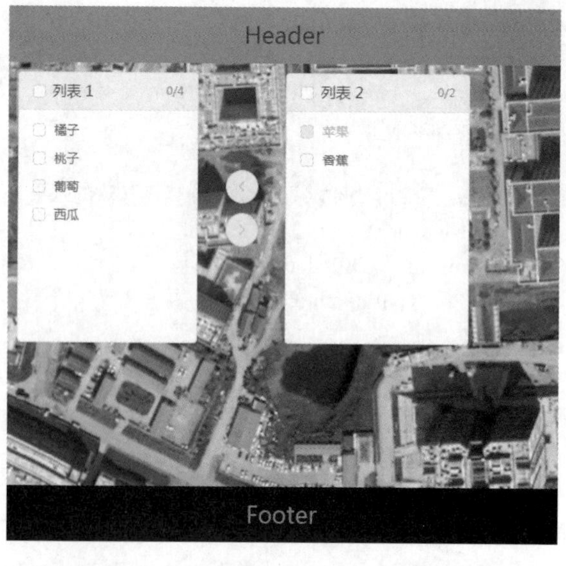

图 3-36　穿梭框数据项属性别名的样式

```
var vm = new Vue({
    el: "#app",
    data: {
        value: [1,2],
        data: [
            { id: 1, name: '苹果', english: "apple", disabled: true },
            { id: 2, name: '香蕉', english: "banana", disabled: false },
            { id: 3, name: '橘子', english: "orange", disabled: false },
            { id: 4, name: '桃子', english: "peach", disabled: false },
            { id: 5, name: '葡萄', english: "garpe", disabled: false },
            { id: 6, name: '西瓜', english: "watermelon", disabled: false }]
    }
})
```

穿梭框数据项属性别名样式对应的标签代码如下：

```
<el-transfer v-model="value" :data="data"  :props="{key: 'id', label: 'name'}">
</el-transfer>
```

3.16 表单（Form）

Element 为表单提供了不同的样式与用法，本节将从典型表单、行内表单、对齐方式、表单验证、自定义验证规则、动态增减表单项、数字类型验证、表单内组件尺寸控制等方面讲解表单的样式与用法。由于表单所占物理空间较大，因此本节分别用一个页面布局来对表单的样式与用法进行讲解。

（1）典型表单。典型表单包括各种表单项，如输入框、选择器、开关、单选框、多选框等。Element 通过一个 el-form 标签和多个 el-form-item 标签的组合实现了典型表单。本示例将结合地图展示典型表单的样式，如图 3-37 所示。

图 3-37 典型表单的样式

本示例首先创建 Vue 实例，data 中的 subjects 是学科选择的数据项，formData 是个对象，该对象的各个值对应于表单中各个组件的 v-model 属性，vectorSource 是一个全局变量，用于显示地图上的文本标注，methods 中的 onSubmit 事件在单击“确定”按钮时被触发；然后结合 OpenLayers 实现地图的文本标注（见图 3-38）。代码如下：

图 3-38　地图文本标注（四）

```
var vm = new Vue({
    el: "#app",
    data:
    {
        subjects: ["学科一", "学科二", "学科三", "学科四"],
        formData: {
            name: "",
            grade: "",
            date: "",
            fresh: false,
            gender: "",
            subjectChecked: [],
            information: ""
        },
        //矢量标注的数据源
        vectorSource: new ol.source.Vector()
    },
    methods: {
        onSubmit:function(){
            //创建矢量标注样式函数
            var createLabelStyle = function (feature) {
                return new ol.style.Style({
                    text: new ol.style.Text({
```

```
                        //位置
                        textAlign: 'center',
                        //基准线
                        textBaseline: 'middle',
                        //文字样式
                        font: 'normal 18px 微软雅黑',
                        //文本内容
                        text: feature.get('name'),
                        //文本填充样式（即文字颜色）
                        fill: new ol.style.Fill({ color: '#aa4400' }),
                        stroke: new ol.style.Stroke({ color: '#ffcc33', width: 2 })
                    })
                });
            }
            //创建点的坐标
            var pt = [2950, 1700];
            //实例化 Vector 要素，通过矢量图层添加到地图容器中
            var iconFeature = new ol.Feature({
                geometry: new ol.geom.Point(pt),
                name:
                "姓名:"+this.formData.name+'\n'+
                "年级:"+this.formData.grade+'\n'+
                "入学日期:"+this.formData.date+'\n'+
                "应届生:"+this.formData.fresh+'\n'+
                "性别:"+this.formData.gender+'\n'+
                "学科选择:"+this.formData.subjectChecked+'\n'+
                "补充:"+this.formData.information
            })
            iconFeature.setStyle(createLabelStyle(iconFeature));
            //添加文本标注之前先清空原来的标注
            this.vectorSource.clear();
            this.vectorSource.addFeature(iconFeature);
            //矢量标注图层
            var vectorLayer = new ol.layer.Vector({
                source: this.vectorSource
            });
            map.addLayer(vectorLayer);
        }
    }
})
```

典型表单样式对应的标签代码如下：

```
<el-form label-width="80px" :model="formData">
    <el-form-item label="姓名">
        <el-input v-model="formData.name" placeholder="请输入姓名"></el-input>
    </el-form-item>
```

```
        <el-form-item label="年级">
            <el-select v-model="formData.grade" placeholder="请选择年级">
                <el-option label="大一" value="大一"></el-option>
                <el-option label="大二" value="大二"></el-option>
                <el-option label="大三" value="大三"></el-option>
                <el-option label="大四" value="大四"></el-option>
            </el-select>
        </el-form-item>
        <el-form-item label="入学日期">
            <el-date-picker type="date" placeholder="选择日期" v-model="formData.date"
                value-format="yyyy-MM-dd HH:mm:ss"></el-date-picker>
        </el-form-item>
        <el-form-item label="应届生">
            <el-switch v-model="formData.fresh"></el-switch>
        </el-form-item>
        <el-form-item label="性别">
            <el-radio-group v-model="formData.gender">
                <el-radio label="男"></el-radio>
                <el-radio label="女"></el-radio>
            </el-radio-group>
        </el-form-item>
        <el-form-item label="学科选择" style="height: 70px;">
            <el-checkbox-group v-model="formData.subjectChecked">
                <el-checkbox v-for="subject of subjects" :label="subject" :key="subject">{{subject}}
                </el-checkbox>
            </el-checkbox-group>
        </el-form-item>
        <el-form-item label="补充" style="height: 40px;">
            <el-input type="textarea" :row="2" v-model="formData.information"></el-input>
        </el-form-item>
        <el-form-item>
            <el-button type="primary" @click="onSubmit">确定</el-button>
            <el-button>取消</el-button>
        </el-form-item>
    </el-form>
```

（2）行内表单。在进行页面布局时，有时在垂直方向的空间会受到限制，如果这时的表单比较简单，则可以对该表单进行横向排版，即行内表单。只需要在 el-form 标签中设置 inline 属性并赋值为 true 即可实现行内表单。

本示例首先创建 Vue 实例，data 中的 formData 是个对象，该对象的各个值对应于表单中各个组件的 v-model 属性，methods 中的 onSubmit 事件在单击“查询”按钮时被触发；然后结合地图展示行内表单的样式（见图 3-39）。代码如下：

图 3-39　行内表单的样式

```
var vm = new Vue({
    el: "#app",
    data: {
        formData: {
            name: "",
            grade: "",
        }
    },
    methods: {
        onSubmit() {
            console.log(this.formData);
        }
    }
})
```

行内表单样式对应的标签代码如下：

```
<el-form label-width="45px" :inline="true" v-model="formData">
    <el-form-item label="姓名">
        <el-input v-model="formData.name" placeholder="请输入姓名" size="mini"></el-input>
    </el-form-item>
    <el-form-item label="年级">
        <el-select v-model="formData.grade" placeholder="请选择年级" size="mini">
            <el-option label="大一" value="大一"></el-option>
            <el-option label="大二" value="大二"></el-option>
            <el-option label="大三" value="大三"></el-option>
            <el-option label="大四" value="大四"></el-option>
```

```
            </el-select>
        </el-form-item>
        <el-form-item>
            <el-button type="primary" @click="onSubmit" size="mini">查询</el-button>
        </el-form-item>
    </el-form>
```

（3）对齐方式。对齐方式是指每行元素的标注所在位置，在 el-form 标签中设置 label-position 属性可以调节标注的位置，该属性的可取值为 right、left 和 top，默认为 rigth。

本示例首先创建 Vue 实例，data 中的 formData 是个对象，该对象的各个值对应于表单中各个组件的 v-model 属性；然后结合地图展示对齐方式的样式（见图 3-40）。代码如下：

图 3-40 对齐方式的样式

```
var vm = new Vue({
    el: "#app",
    data: {
        formData: {
            name: "",
            grade: "",
        }
    },
})
```

对齐方式样式对应的标签代码如下：

```
<el-form   v-model="formData"   label-width="80px" label-position="top">
    <el-form-item label="姓名">
        <el-input v-model="formData.name" placeholder="请输入姓名" size="mini"></el-input>
```

```
        </el-form-item>
        <el-form-item label="入学日期">
            <el-date-picker type="date" placeholder="选择日期" v-model="formData.date"></el-date-picker>
        </el-form-item>
    </el-form>
```

（4）表单验证。表单验证是指在防止用户犯错的前提下，尽可能让用户更早地发现并纠正错误。Element 提供了表单验证功能，只需要通过 rules 属性传入约定的验证规则，并将 Form-Item 的 prop 属性设置为需要验证的字段名即可。

本示例首先创建 Vue 实例，data 中的 subjects 是学科选择的数据项，formData 是个对象，该对象的各个值对应于表单中各个组件的 v-model 属性，rules 是个对象，该对象为表单各个组件的验证规则，methods 中的 submitForm 事件在单击“立即创建”按钮时被触发，resetForm 事件在单击“重置”按钮时被触发；然后结合地图展示表单验证的样式（见图 3-41）。代码如下：

图 3-41　表单验证的样式

```
var vm = new Vue({
    el: "#app",
    data: {
        subjects: ["学科一", "学科二", "学科三", "学科四"],
        ruleForm: {
            name: "",
            grade: "",
            date: "",
            fresh: false,
            gender: "",
            subjectChecked: [],
            information: ""
        },
        rules: {
```

```
                name: [
                    { required: true, message: '请输入姓名', trigger: 'blur' },
                    { min: 2, max: 5, message: '长度在 2 到 5 个字符', trigger: 'blur' }
                ],
                grade: [
                    { required: true, message: '请选择年级', trigger: 'change' }
                ],
                date: [
                    { type: 'date', required: true, message: '请选择日期', trigger: 'change' }
                ],
                fresh: [{ required: true }],
                gender: [
                    { required: true, message: '请选择性别', trigger: 'change' }
                ],
                subjectChecked: [
                    { type: 'array', required: true, message: '请至少选择一个学科', trigger: 'change' }
                ],
                information: [
                    { required: true, message: '请填写补充说明', trigger: 'blur' }
                ]
            }
        },
        methods: {
            submitForm(formName) {
                alert(formName)
                this.$refs[formName].validate(function (valid) {
                    if (valid) {
                        alert('submit!');
                    } else {
                        return false;
                    }
                });
            },
            resetForm(formName) {
                this.$refs[formName].resetFields();
            }
        }
    })
```

表单验证样式对应的标签代码如下：

```
<el-form :model="ruleForm" :rules="rules" ref="ruleForm" label-width="100px" class="demo-ruleForm">
    <el-form-item label="姓名" prop="name">
        <el-input v-model="ruleForm.name"></el-input>
    </el-form-item>
    <el-form-item label="年级" prop="grade">
        <el-select v-model="ruleForm.grade" placeholder="请选择年级">
```

```
                <el-option label="大一" value="大一"></el-option>
                <el-option label="大二" value="大二"></el-option>
                <el-option label="大三" value="大三"></el-option>
                <el-option label="大四" value="大四"></el-option>
            </el-select>
        </el-form-item>
        <el-form-item label="入学日期" prop="date">
            <el-date-picker type="date" placeholder="选择日期" v-model="ruleForm.date"
                value-format="yyyy-MM-dd HH:mm:ss"></el-date-picker>
        </el-form-item>
        <el-form-item label="应届生" prop="fresh">
            <el-switch v-model="ruleForm.fresh"></el-switch>
        </el-form-item>
        <el-form-item label="性别" prop="gender">
            <el-radio-group v-model="ruleForm.gender">
                <el-radio label="男"></el-radio>
                <el-radio label="女"></el-radio>
            </el-radio-group>
        </el-form-item>
        <el-form-item label="学科选择" style="height: 90px;" prop="subjectChecked">
            <el-checkbox-group v-model="ruleForm.subjectChecked">
                <el-checkbox v-for="subject of subjects" :label="subject" :key="subject">{{subject}}
                </el-checkbox>
            </el-checkbox-group>
        </el-form-item>
        <el-form-item label="补充" prop="information" style="height: 50px;">
            <el-input type="textarea" v-model="ruleForm.information"></el-input>
        </el-form-item>
        <el-form-item>
            <el-button type="primary" @click="submitForm('ruleForm')">立即创建</el-button>
            <el-button @click="resetForm('ruleForm')">重置</el-button>
        </el-form-item>
    </el-form>
```

（5）自定义验证规则。在表单验证中，有时需要自定义一些验证规则。在 rules 对象中为每个子对象设置 validator 属性，该属性所对应的变量可以用函数的形式来自定义验证规则。

本示例首先创建 Vue 实例，此时的 data 为一个函数，data 内的 checkAge、validatePass 和 validatePass2 分别为名字、密码和确认密码输入框的自定义规则函数，ruleForm 是个对象，该对象的各个值对应于表单中各个组件的 v-model 属性，rules 是个对象，该对象为表单各个组件的验证规则，methods 中的 submitForm 事件在单击“提交”按钮时被触发，resetForm 事件在单击“重置”按钮时被触发；然后结合地图展示自定义验证规则的样式（见图 3-42）。代码如下：

```
var vm = new Vue({
    el: "#app",
    data: function () {
```

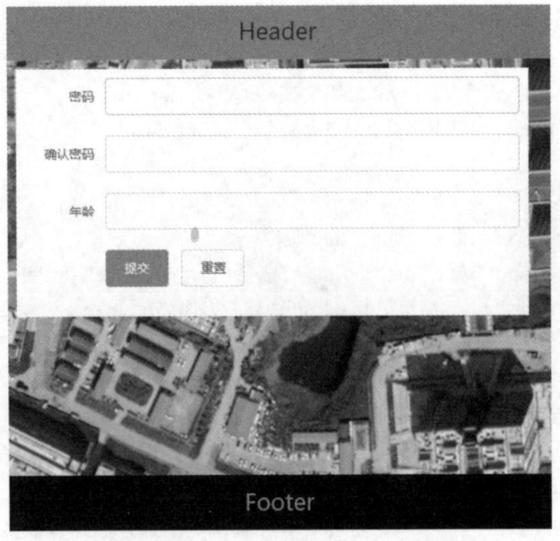

图 3-42　自定义验证规则的样式

```
//自定义年龄输入验证规则
var checkAge = function (rule, value, callback) {
    if (!value) {
        return callback(new Error('年龄不能为空'));
    }
    if (!Number.isInteger(value)) {
        callback(new Error('请输入数字值'));
    }
    else if (value < 18) {
        callback(new Error('必须年满 18 岁'));
    }
    else {
        callback();
    }
};
//自定义密码输入验证规则
var validatePass = function (rule, value, callback) {
    if (value === '') {
        callback(new Error('请输入密码'));
    }
    else if (this.ruleForm.checkPass !== '') {
        this.$refs.ruleForm.validateField('checkPass');
        callback();
    }
};
 //自定义确认密码输入验证规则
var validatePass2 = function (rule, value, callback) {
    if (value === '') {
        callback(new Error('请再次输入密码'));
    }
```

```
            else if (value !== this.ruleForm.pass) {
                callback(new Error('两次输入密码不一致!'));
            }
            else {
                callback();
            }
        };
        return {
            //表单数据对象
            ruleForm: {
                pass: '',
                checkPass: '',
                age: ''
            },
            //规则对象
            rules: {
                pass: [
                    { validator: validatePass, trigger: 'blur' }
                ],
                checkPass: [
                    { validator: validatePass2, trigger: 'blur' }
                ],
                age: [
                    { validator: checkAge, trigger: 'blur' }
                ]
            }
        }

    },
    methods: {
        //单击“提交”按钮时的响应事件
        submitForm(formName) {
            this.$refs[formName].validate((valid) => {
                if (valid) {
                    alert('submit!');
                } else {
                    console.log('error submit!!');
                    return false;
                }
            });
        },
        //单击“重置”按钮时的响应事件
        resetForm(formName) {
            this.$refs[formName].resetFields();
        }
    },
```

```
})
```

自定义验证规则样式对应的标签代码如下：

```
<el-form :model="ruleForm" status-icon :rules="rules" ref="ruleForm" label-width="100px"
    class="demo-ruleForm">
    <el-form-item label="密码" prop="pass">
        <el-input type="password" v-model="ruleForm.pass" autocomplete="off"></el-input>
    </el-form-item>
    <el-form-item label="确认密码" prop="checkPass">
        <el-input type="password" v-model="ruleForm.checkPass" autocomplete="off"></el-input>
    </el-form-item>
    <el-form-item label="年龄" prop="age">
        <el-input v-model.number="ruleForm.age"></el-input>
    </el-form-item>
    <el-form-item>
        <el-button type="primary" @click="submitForm('ruleForm')">提交</el-button>
        <el-button @click="resetForm('ruleForm')">重置</el-button>
    </el-form-item>
</el-form>
```

（6）动态增减表单项。在 Element 提供的表单中，可以动态地增减表单项。动态增减表单项的方法是，首先选择动态增减组件，该组件是通过 v-for 对列表进行渲染得到的；然后为“删除”按钮绑定一个用于删除组件的事件，为“新增域名”按钮绑定一个用于添加组件的事件。

本示例首先创建 Vue 实例，data 中的 dynamicValidateForm 为动态渲染表单组件的数据对象，methods 中的 submitForm 为单击“提交”按钮时的响应事件，resetForm 为单击“重置”按钮时的响应事件，removeDomain 为单击“删除”按钮时的响应事件，addDomain 为单击“新增域名”按钮时的响应事件；然后结合地图展示动态增减表单项的样式（见图 3-43）。代码如下：

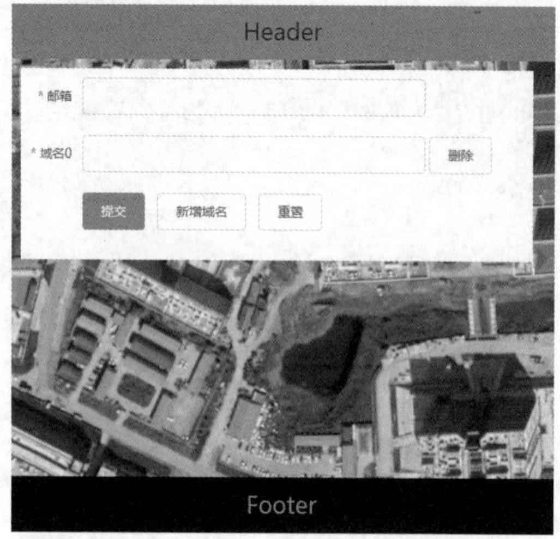

图 3-43　动态增减表单项的样式

```
var vm = new Vue({
    el: "#app",
    data: {
        //动态渲染组件的数据对象
        dynamicValidateForm: {
            domains: [{
                value: ''
            }],
            email: ''
        }
    },
    methods: {
        //单击“提交”按钮时的响应事件
        submitForm(formName) {
            this.$refs[formName].validate((valid) => {
                if (valid) {
                    alert('submit!');
                } else {
                    console.log('error submit!!');
                    return false;
                }
            });
        },
        //单击“重置”按钮时的响应事件
        resetForm(formName) {
            this.$refs[formName].resetFields();
        },
        //单击“删除”按钮时的响应事件
        removeDomain(item) {
            var index = this.dynamicValidateForm.domains.indexOf(item)
            if (index !== -1) {
                this.dynamicValidateForm.domains.splice(index, 1)
            }
        },
        //单击“新增域名”按钮时的响应事件
        addDomain() {
            this.dynamicValidateForm.domains.push({
                value: '',
                key: Date.now()
            });
        }
    },
})
```

动态增减表单项样式对应的标签代码如下：

```
<el-form :model="dynamicValidateForm" ref="dynamicValidateForm" label-width="70px">
```

```
        <el-form-item prop="email" label="邮箱" :rules="[
            { required: true, message: '请输入邮箱地址', trigger: 'blur' },
            { type: 'email', message: '请输入正确的邮箱地址', trigger: ['blur', 'change'] }
          ]">
            <el-input v-model="dynamicValidateForm.email"></el-input>
        </el-form-item>
        <el-form-item v-for="(domain, index) in dynamicValidateForm.domains" :label="'域名' + index"
            :key="domain.key" :prop="'domains.' + index + '.value'" :rules="{
            required: true, message: '域名不能为空', trigger: 'blur'
          }">
            <el-input v-model="domain.value"></el-input>
            <el-button @click.prevent="removeDomain(domain)">删除</el-button>
        </el-form-item>
        <el-form-item>
            <el-button type="primary" @click="submitForm('dynamicValidateForm')">提交</el-button>
            <el-button @click="addDomain">新增域名</el-button>
            <el-button @click="resetForm('dynamicValidateForm')">重置</el-button>
        </el-form-item>
    </el-form>
```

（7）数字类型验证。在表单的输入框中，有时需要限制其输入类型，最典型的就是只允许输入数字。想要实现数字类型验证，只需要在 el-input 标签中的 v-model 属性上添加.number 修饰符即可。

本示例首先创建 Vue 实例，data 中的 numberValidateForm 是个对象，该对象的 age 对应于表单中输入框的 v-model 属性，methods 中的 submitForm 为单击“提交”按钮时的响应事件，resetForm 为单击“重置”按钮时的响应事件；然后结合地图展示数字类型验证的样式（见图 3-44）。代码如下：

图 3-44　数字类型验证的样式

```
var vm = new Vue({
    el: "#app",
    data: {
        numberValidateForm: {
            age: ''
        }
    },
    methods: {
        //单击“提交”按钮时的响应事件
        submitForm(formName) {
            this.$refs[formName].validate((valid) => {
                if (valid) {
                    alert('submit!');
                } else {
                    console.log('error submit!!');
                    return false;
                }
            });
        },
        //单击“重置”按钮时的响应事件
        resetForm(formName) {
            this.$refs[formName].resetFields();
        }
    },
})
```

数字类型验证样式对应的标签代码如下：

```
<el-form :model="numberValidateForm" ref="numberValidateForm" label-width="100px" class="demo-ruleForm">
    <el-form-item label="年龄" prop="age" :rules="[
            { required: true, message: '年龄不能为空'},
            { type: 'number', message: '年龄必须为数字'}
        ]">
        <el-input type="age" v-model.number="numberValidateForm.age" autocomplete="off"></el-input>
    </el-form-item>
    <el-form-item>
        <el-button type="primary" @click="submitForm('numberValidateForm')">提交</el-button>
        <el-button @click="resetForm('numberValidateForm')">重置</el-button>
    </el-form-item>
</el-form>
```

（8）表单内组件尺寸控制。在 el-form 标签中设置 size 属性可以控制表单的尺寸。el-form-item 标签也可以通过 size 属性来控制表单的尺寸，size 的取值为 large、medium、small、mini，默认为 large。在典型表单的基础上，将 el-form 标签中的 size 属性设置为 mini 可以控制表单的尺寸，表单内组件尺寸控制的样式如图 3-4 所示。

图 3-45　表单内组件尺寸控制的样式

只需要将典型表单中 el-form 标签中的 size 属性设置为 mini 即可实现图 3-45 所示的样式，为了避免代码冗余，该示例不再展示代码。

3.17　思考与练习题

通过本章介绍的组件，实现一个学生管理的表单。

第4章 数据组件

Element 为数据组件提供了非常丰富、详细的样式，本章将结合 Vue 和 OpenLayers 详细地介绍 Element 的数据组件，主要包括表格、标签、进度条、树形组件、分页、标注、头像等的样式和用法。

4.1 表格（Table）

Element 为表格提供了不同的样式与用法，本节从基础表格、带斑马纹表格、带边框表格、带状态表格、固定表头、固定列、固定列和表头、流体高度、多级表头、单选、多选、排序、筛选、自定义列模板、展开行、树形数据与懒加载、自定义表头、表尾合计行、合并行或列、自定义索引等方面讲解表格的样式和用法。

由于表格所占的物理空间较大，因此本节分别用一个页面布局对每个样式进行讲解。

（1）基础表格。Element 通过对 el-table 的封装提供了表格，其中 el-table 标签中的 data 属性所对应的对象数组是表格中显示的数据，el-table 标签中的子标签 el-table-column 中的 prop 属性对应于列内容的字段名，label 属性用来定义表格的列名，可以使用 width 属性来定义列宽。

本示例将结合地图展示表格的样式，其中基础表格的样式如图 4-1 所示。

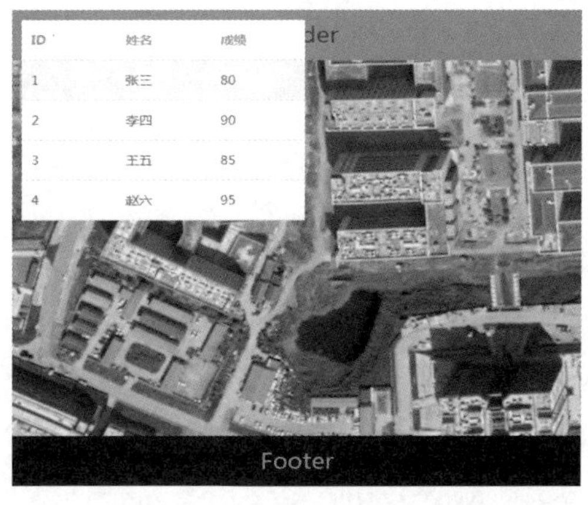

图 4-1　基础表格的样式

本示例首先创建 Vue 实例，data 中的 tableData 是一个对象数组，每个对象都表示表格中的一行数据，vectorSource 是一个全局变量，用于显示地图上的文本标注，methods 中的 handleRowClick 事件为表格的行单击事件；然后结合 OpenLayers 实现地图的文本标注（见图 4-2）。代码如下：

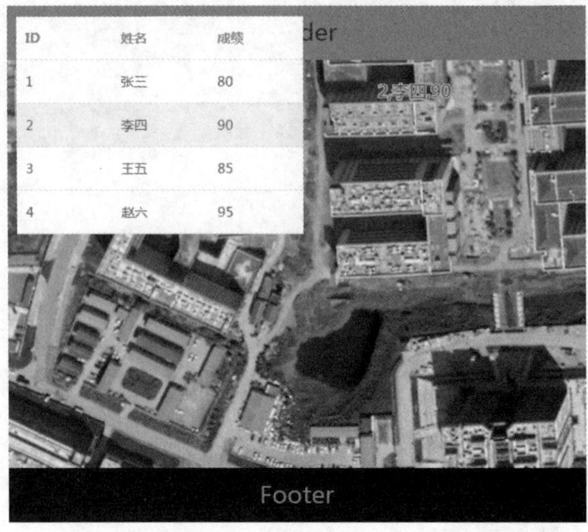

图 4-2　地图文本标注

```
var vm = new Vue({
    el: "#app",
    data:
    {
        tableData:[{
            ID:1,
            name:"张三",
            grade:80

        },{
            ID:2,
            name:"李四",
            grade:90
        },{
            ID:3,
            name:"王五",
            grade:85
        },{
            ID:4,
            name:"赵六",
            grade:95
        }],
        //矢量标注的数据源
        vectorSource: new ol.source.Vector()
```

```
    },
    methods: {
        handleRowClick: function (val) {
            console.log(val.ID)
            //创建矢量标注样式函数
            var createLabelStyle = function (feature) {
                return new ol.style.Style({
                    text: new ol.style.Text({
                        //位置
                        textAlign: 'center',
                        //基准线
                        textBaseline: 'middle',
                        //文字样式
                        font: 'normal 18px 微软雅黑',
                        //文本内容
                        text: feature.get('name'),
                        //文本填充样式（即文字颜色）
                        fill: new ol.style.Fill({ color: '#aa4400' }),
                        stroke: new ol.style.Stroke({ color: '#ffcc33', width: 2 })
                    })
                });
            }
            //创建点的坐标
            var pt = [2750, 1900];
            //实例化 Vector 要素，通过矢量图层添加到地图容器中
            var iconFeature = new ol.Feature({
                geometry: new ol.geom.Point(pt),
                name:val.ID+","+val.name+","+val.grade
            })
            iconFeature.setStyle(createLabelStyle(iconFeature));
            //添加文本标注之前先清空原来的标注
            this.vectorSource.clear();
            this.vectorSource.addFeature(iconFeature);
            //矢量标注图层
            var vectorLayer = new ol.layer.Vector({
                source: this.vectorSource
            });
            map.addLayer(vectorLayer);
        }
    }
})
```

基础表格样式对应的标签代码如下：

```
<el-table :data="tableData" @row-click="handleRowClick" >
    <el-table-column prop="ID" label="ID" width="100">
    </el-table-column>
```

```
        <el-table-column prop="name" label="姓名" width="100">
        </el-table-column>
        <el-table-column prop="grade" label="成绩" width="100">
        </el-table-column>
</el-table>
```

（2）带斑马纹表格。使用带斑马纹的表格可以更容易区分出不同行的数据，只需要在基础表格的 el-table 标签中添加 stripe 属性即可实现带斑马纹表格。

本示例首先创建 Vue 实例，data 中的 tableData 是一个对象数组，每个对象都表示表格中的每一行数据；然后结合地图展示带斑马纹表格的样式，如图 4-3 所示。代码如下：

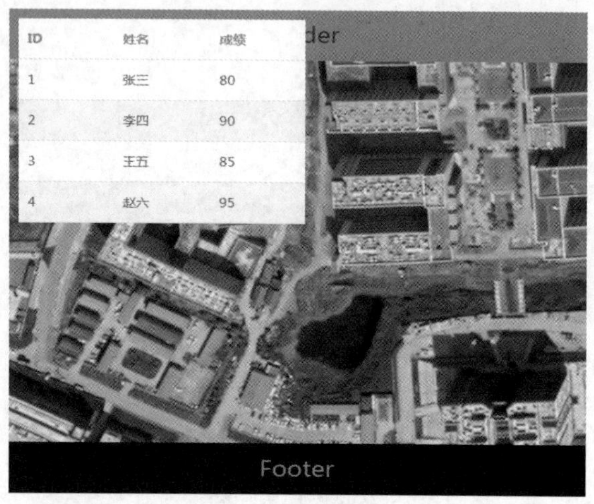

图 4-3　带斑马纹表格的样式

```
var vm = new Vue({
    el: "#app",
    data:
    {
        tableData:[{
            ID:1,
            name:"张三",
            grade:80
        },{
            ID:2,
            name:"李四",
            grade:90
        },{
            ID:3,
            name:"王五",
            grade:85
        },{
            ID:4,
            name:"赵六",
            grade:95
```

```
        }],
    }
})
```

带斑马纹表格样式对应的标签代码如下：

```
<el-table :data="tableData" stripe >
    <el-table-column prop="ID" label="ID" width="100">
    </el-table-column>
    <el-table-column prop="name" label="姓名" width="100">
    </el-table-column>
    <el-table-column prop="grade" label="成绩" width="100">
    </el-table-column>
</el-table>
```

（3）带边框表格。在默认情况下，表格没有垂直方向的边框。在 el-table 标签中添加 border 属性即可使表格带有垂直方向的边框。

本示例首先创建 Vue 实例，该实例与带斑马纹表格的 Vue 实例一样，故不再展示 Vue 实例的代码；然后结合地图展示带边框表格的样式，如图 4-4 所示。

图 4-4　带边框表格的样式

带边框表格样式对应的标签代码如下：

```
<el-table :data="tableData" border >
    <el-table-column prop="ID" label="ID" width="100">
    </el-table-column>
    <el-table-column prop="name" label="姓名" width="100">
    </el-table-column>
    <el-table-column prop="grade" label="成绩" width="100">
    </el-table-column>
</el-table>
```

（4）带状态表格。带状态表格可以高亮显示表格中的内容，便于区分成功、信息、警告、危险等信息。

本示例首先创建 Vue 实例，data 中的 tableData 是一个对象数组，每个对象都表示表格中的每一行数据，methods 中的 tableRowClassName 方法用于接收一个参数，该参数是一个对象，里面两个键，其中一个是 rowIndex，表示表格每行的索引号，根据索引号可以为该行返回不同的样式，warning-row 和 success-row 是自定义的 CSS 样式；然后结合地图展示带状态表格的样式，如图 4-5 所示。代码如下：

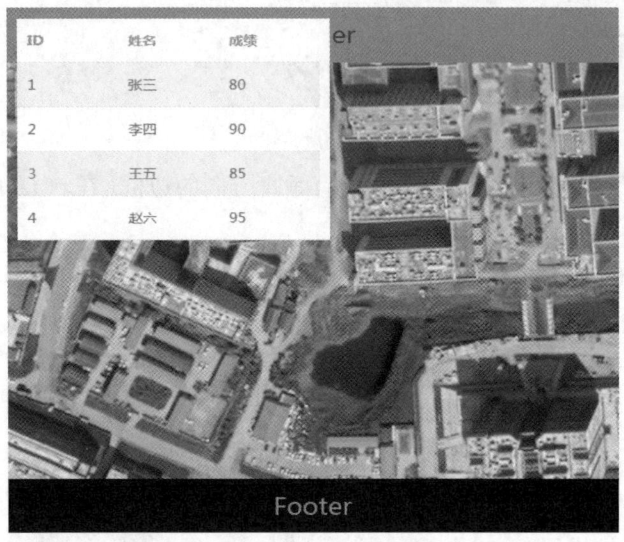

图 4-5　带状态表格的样式

```
var vm = new Vue({
    el: "#app",
    data:
    {
        tableData: [{
            ID: 1,
            name: "张三",
            grade: 80
        }, {
            ID: 2,
            name: "李四",
            grade: 90
        }, {
            ID: 3,
            name: "王五",
            grade: 85
        }, {
            ID: 4,
            name: "赵六",
            grade: 95
```

```
            }],
        },
        methods: {
            tableRowClassName: function ({ row, rowIndex }) {
                if (rowIndex === 0) {
                    return 'warning-row';
                }
                else if (rowIndex === 2) {
                    return 'success-row';
                }
                return ' ';
            }
        },
    })
```

带状态表格样式对应的标签代码如下：

```
<el-table :data="tableData" :row-class-name="tableRowClassName">
    <el-table-column prop="ID" label="ID" width="100">
    </el-table-column>
    <el-table-column prop="name" label="姓名" width="100">
    </el-table-column>
    <el-table-column prop="grade" label="成绩" width="100">
    </el-table-column>
</el-table>
```

（5）固定表头。固定表头是表格的一种常见用法，当纵向内容过多时，可以使用固定表头。只需要在 el-table 元素中定义 height 属性即可实现固定表头。

本示例首先创建 Vue 实例，data 中的 tableData 是一个对象数组，每个对象都表示表格中的每一行数据；然后结合地图展示固定表头的样式，如图 4-5 所示。

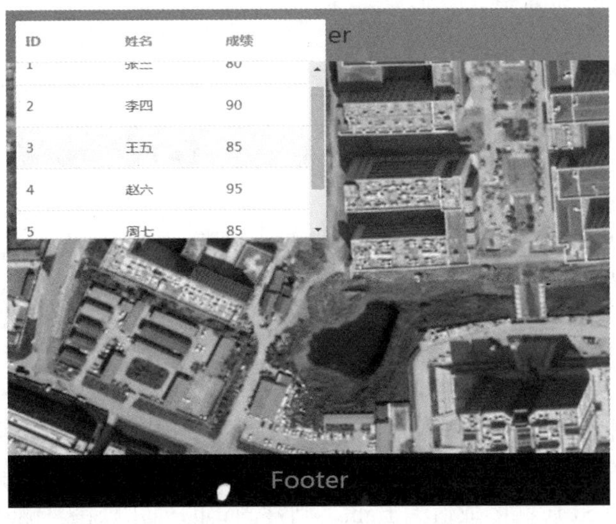

图 4-6　固定表头的样式

```
var vm = new Vue({
    el: "#app",
    data:
    {
        tableData:[{
            ID:1,
            name:"张三",
            grade:80
        },{
            ID:2,
            name:"李四",
            grade:90
        },{
            ID:3,
            name:"王五",
            grade:85
        },{
            ID:4,
            name:"赵六",
            grade:95
        },{
            ID:5,
            name:"周七",
            grade:85
        },{
            ID:6,
            name:"吴八",
            grade:90
        }],
    }
}})
```

固定表头样式对应的标签代码如下：

```
<el-table :data="tableData" height="250">
    <el-table-column prop="ID" label="ID" width="100">
    </el-table-column>
    <el-table-column prop="name" label="姓名" width="100">
    </el-table-column>
    <el-table-column prop="grade" label="成绩" width="100">
    </el-table-column>
</el-table>
```

（6）固定列。固定列是表格的一种常见用法，当横向内容过多时，可以使用固定列。只需要在待固定的列中添加 fixed 属性即可实现固定列。若 fixed 属性的取值为 left，则被固定的列在表格左侧；若 fixed 属性的取值为 right，则被固定的列在表格右侧。fixed 属性的默认取值为 left。

本示例首先创建 Vue 实例，data 中的 tableData 是一个对象数组，每个对象都表示表格中的每一行数据；然后结合地图展示固定列的样式，如图 4-7 所示。代码如下：

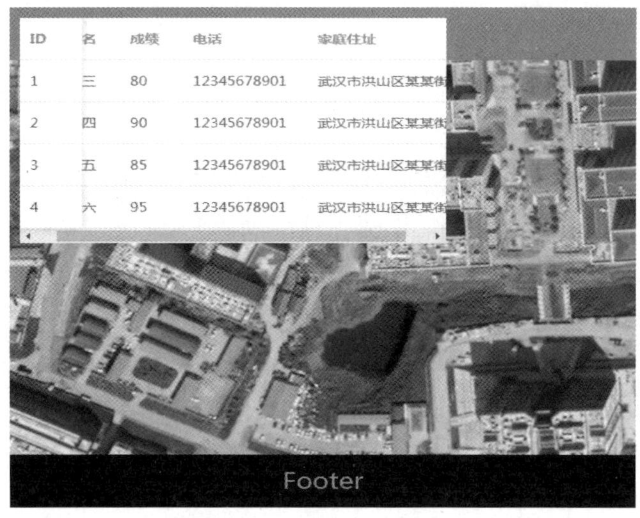

图 4-7　固定列的样式

```
var vm = new Vue({
    el: "#app",
    data:
    {
        tableData: [{
            ID: 1,
            name: "张三",
            grade: 80,
            phone: "12345678901",
            adress: "武汉市洪山区某某街道"
        }, {
            ID: 2,
            name: "李四",
            grade: 90,
            phone: "12345678901",
            adress: "武汉市洪山区某某街道"
        }, {
            ID: 3,
            name: "王五",
            grade: 85,
            phone: "12345678901",
            adress: "武汉市洪山区某某街道"
        }, {
            ID: 4,
            name: "赵六",
            grade: 95,
            phone: "12345678901",
```

```
                adress: "武汉市洪山区某某街道"
            }],
        }
})
```

固定列样式对应的标签代码如下：

```
<el-table :data="tableData">
    <el-table-column prop="ID" label="ID" width="60" fixed>
    </el-table-column>
    <el-table-column prop="name" label="姓名" width="60">
    </el-table-column>
    <el-table-column prop="grade" label="成绩" width="60">
    </el-table-column>
    <el-table-column prop="phone" label="电话" width="120">
    </el-table-column>
    <el-table-column prop="adress" label="家庭住址" width="160">
    </el-table-column>
</el-table>
```

（7）固定列和表头。固定列和表头也是表格的一种常见用法，当横/纵向内容过多时，可以使用固定列和表头。在使用固定列和表头时，只需要将固定表头和固定列结合起来即可。

本示例首先创建 Vue 实例，data 中的 tableData 是一个对象数组，每个对象都表示表格中的每一行数据；然后结合地图展示固定列和表头的样式，如图 4-8 所示。代码如下：

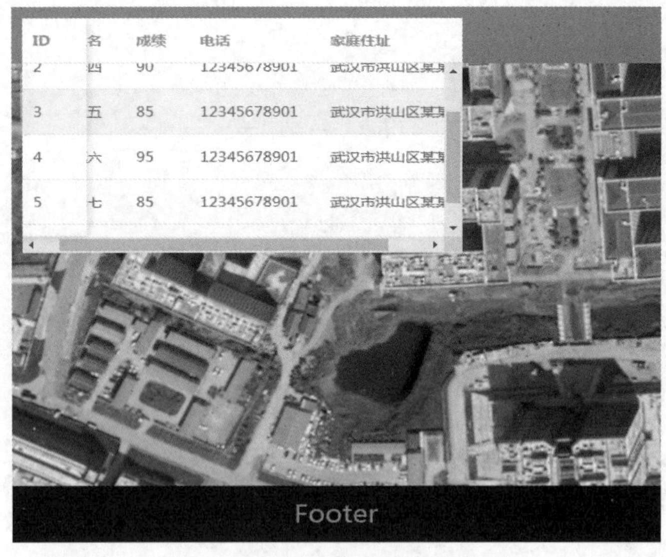

图 4-8　固定列和表头的样式

```
var vm = new Vue({
    el: "#app",
    data:
    {
        tableData: [{
```

```
            ID: 1,
            name: "张三",
            grade: 80,
            phone: "12345678901",
            adress: "武汉市洪山区某某街道"
        }, {
            ID: 2,
            name: "李四",
            grade: 90,
            phone: "12345678901",
            adress: "武汉市洪山区某某街道"
        }, {
            ID: 3,
            name: "王五",
            grade: 85,
            phone: "12345678901",
            adress: "武汉市洪山区某某街道"
        }, {
            ID: 4,
            name: "赵六",
            grade: 95,
            phone: "12345678901",
            adress: "武汉市洪山区某某街道"
        },{
            ID:5,
            name:"周七",
            grade:85,
            phone: "12345678901",
            adress: "武汉市洪山区某某街道"
        },{
            ID:6,
            name:"吴八",
            grade:90,
            phone: "12345678901",
            adress: "武汉市洪山区某某街道"
        }],
    }
})
```

固定列和表头样式对应的标签代码如下：

```
<el-table :data="tableData" height="250">
    <el-table-column prop="ID" label="ID" width="60" fixed="left" >
    </el-table-column>
    <el-table-column prop="name" label="姓名" width="60">
    </el-table-column>
    <el-table-column prop="grade" label="成绩" width="60">
```

```
        </el-table-column>
        <el-table-column prop="phone" label="电话" width="120">
        </el-table-column>
        <el-table-column prop="adress" label="家庭住址" width="160">
        </el-table-column>
</el-table>
```

（8）流体高度。当表格的数据动态变化时，即可以移除和新增表格的行数据时，可以为该表格设置一个最大高度。若表格的高度大于设置的最大高度，则会显示一个滚动条。

本示例首先创建 Vue 实例，data 中的 tableData 是一个对象数组，每个对象都表示表格中的每一行数据，methods 中的 deleteRow 是单击表格中某行时的响应事件；然后结合地图展示流体高度的样式，如图 4-9 所示。

图 4-9　流体高度的样式

```
var vm = new Vue({
    el: "#app",
    data:
    {
        tableData: [{
            ID: 1,
            name: "张三",
            grade: 80,
            phone: "12345678901",
            adress: "武汉市洪山区某某街道"
        }, {
            ID: 2,
            name: "李四",
            grade: 90,
```

```
                phone: "12345678901",
                adress: "武汉市洪山区某某街道"
            }, {
                ID: 3,
                name: "王五",
                grade: 85,
                phone: "12345678901",
                adress: "武汉市洪山区某某街道"
            }, {
                ID: 4,
                name: "赵六",
                grade: 95,
                phone: "12345678901",
                adress: "武汉市洪山区某某街道"
            }, {
                ID: 5,
                name: "周七",
                grade: 85,
                phone: "12345678901",
                adress: "武汉市洪山区某某街道"
            }, {
                ID: 6,
                name: "吴八",
                grade: 90,
                phone: "12345678901",
                adress: "武汉市洪山区某某街道"
            }],
    },
    methods: {
        //单击表格中某行时的响应事件
        deleteRow(index, rows) {
            //rows 是表格数据，以数组的形式存储，通过 splice 和当前行的索引号可移除数据
            rows.splice(index, 1);
        }
    }
})
```

流体高度样式对应的标签代码如下：

```
<el-table :data="tableData" height="250">
    <el-table-column prop="ID" label="ID" width="60" fixed="left">
    </el-table-column>
    <el-table-column prop="name" label="姓名" width="60">
    </el-table-column>
    <el-table-column prop="grade" label="成绩" width="60">
    </el-table-column>
    <el-table-column prop="phone" label="电话" width="120">
```

```
        </el-table-column>
        <el-table-column prop="adress" label="家庭住址" width="160">
        </el-table-column>
        <el-table-column fixed="right" label="操作" width="80">
            <!-- 创建“移除”按钮的模板，通过 slot-scope 可以获取到表格内部数据-->
            <template slot-scope="scope">
                <!-- scope.$index 为当前行的索引 -->
                <el-button @click="deleteRow(scope.$index, tableData)" type="text"
                    size="small">
                    移除
                </el-button>
            </template>
        </el-table-column>
    </el-table>
```

（9）多级表头。当所要呈现的数据结构比较复杂时，可以使用多级表头来展现数据的层次关系。在使用多级表头时，只需要在 el-table-colunn 里面嵌套 el-table-column 即可。

本示例首先创建 Vue 实例，该 Vue 实例与固定列和表头的 Vue 实例一样，故不再展示 Vue 实例的代码；然后结合地图展示多级表头的样式，如图 4-10 所示。

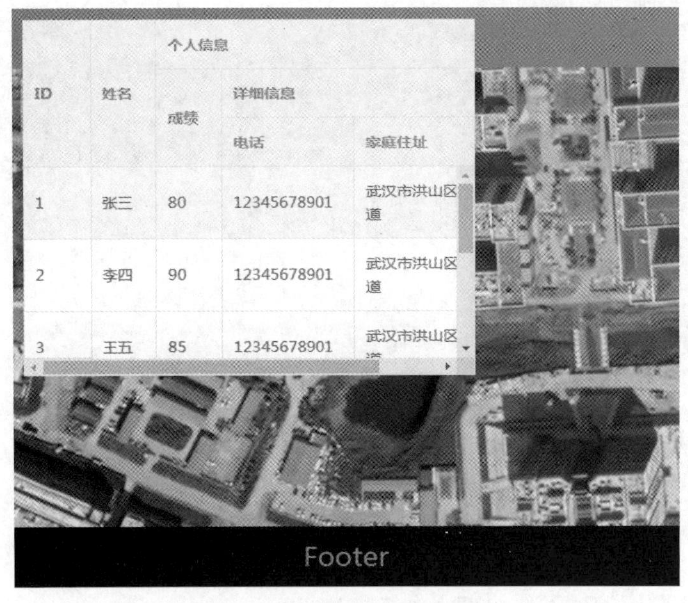

图 4-10　多级表头的样式

多级表头样式对应的标签代码如下：

```
<el-table :data="tableData" height="350">
    <el-table-column prop="ID" label="ID" width="60" fixed="left">
    </el-table-column>
    <el-table-column prop="name" label="姓名" width="60">
    </el-table-column>
    <el-table-column label="个人信息">
```

```
            <el-table-column prop="grade" label="成绩" width="60">
            </el-table-column>
            <el-table-column label="详细信息">
                <el-table-column prop="phone" label="电话" width="120">
                </el-table-column>
                <el-table-column prop="adress" label="家庭住址" width="160">
                </el-table-column>
            </el-table-column>
        </el-table-column>
    </el-table>
```

（10）单选。单选是指单击表格的某一行时，该行会变色。只需要在 el-table 标签中添加 highlight-current-row 属性即可实现单选，current-change 事件用来管理选中某行时的触发事件，它会传入 currentRow 和 oldCurrentRow。如果需要显示索引，则可以增加一列 el-table-column，设置 type 属性为 index 即可显示从 1 开始的索引号。

本示例首先创建 Vue 实例，data 中的 tableData 是一个对象数组，每个对象都表示表格中的一行数据，methods 中的 handleCurrentChange 是单击表格中某行时的响应事件，该方法有两个参数，第一个参数 currentRow 为当前行的数据，第二个参数 oldRow 为上一次单击行的数据；然后结合地图展示单选的样式，如图 4-11 所示。代码如下：

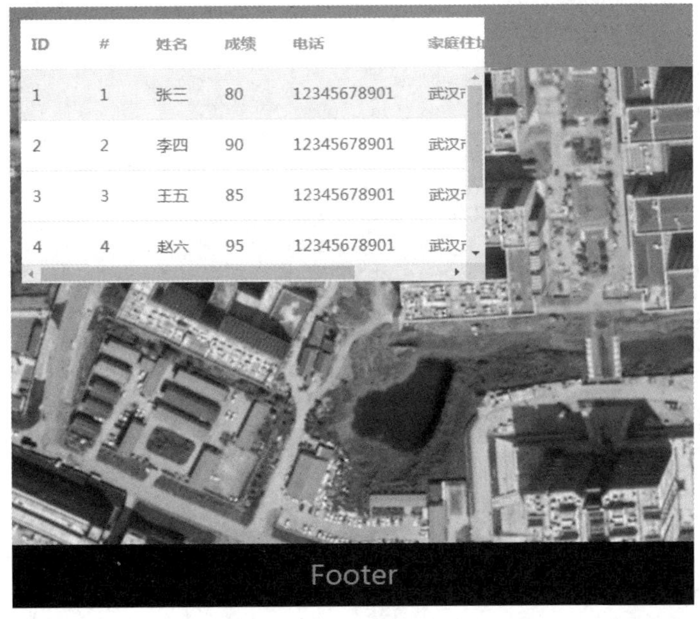

图 4-11　单选的样式

```
var vm = new Vue({
    el: "#app",
    data:
    {
        tableData: [{
            ID: 1,
```

```
                name: "张三",
                grade: 80,
                phone: "12345678901",
                adress: "武汉市洪山区某某街道"
            }, {
                ID: 2,
                name: "李四",
                grade: 90,
                phone: "12345678901",
                adress: "武汉市洪山区某某街道"
            }, {
                ID: 3,
                name: "王五",
                grade: 85,
                phone: "12345678901",
                adress: "武汉市洪山区某某街道"
            }, {
                ID: 4,
                name: "赵六",
                grade: 95,
                phone: "12345678901",
                adress: "武汉市洪山区某某街道"
            }, {
                ID: 5,
                name: "周七",
                grade: 85,
                phone: "12345678901",
                adress: "武汉市洪山区某某街道"
            }, {
                ID: 6,
                name: "吴八",
                grade: 90,
                phone: "12345678901",
                adress: "武汉市洪山区某某街道"
            }],
        },
        methods:{
            //单击表格某行时的响应事件
            handleCurrentChange:function(currentRow,oldRow){
                console.log(currentRow);
                console.log(oldRow);
            }
        }
    })
```

单选样式对应的标签代码如下：

```
<el-table :data="tableData" height="250" highlight-current-row   @current-change="handleCurrentChange">
    <el-table-column type="index" width="50">
    </el-table-column>
    <el-table-column prop="ID" label="ID" width="60" fixed="left">
    </el-table-column>
    <el-table-column prop="name" label="姓名" width="60">
    </el-table-column>
    <el-table-column prop="grade" label="成绩" width="60">
    </el-table-column>
    <el-table-column prop="phone" label="电话" width="120">
    </el-table-column>
    <el-table-column prop="adress" label="家庭住址" width="160">
    </el-table-column>
</el-table>
```

（11）多选。表格的多选是使用多选框的形式来表示的，Element 实现多选的方法非常简单，在 el-table 标签中添加一个 el-table-column 标签，将该标签中的 type 属性设置为 selection 即可。

本示例首先创建 Vue 实例，data 中的 tableData 是一个对象数组，每个对象都表示表格中的一行数据，multipleSelection 为一空数组，在多选框被勾选时，用于存放所选数据，methods 中的 handleSelectionChange 是单击表格的多选框时的响应事件，该方法会接收一个参数，该参数是当前所选数据；然后结合地图展示多选的样式，如图 4-12 所示。代码如下：

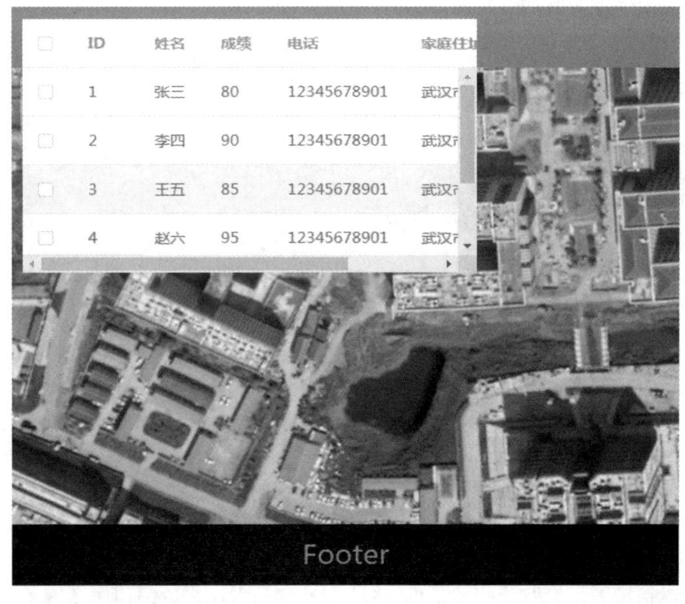

图 4-12　多选的样式

```
var vm = new Vue({
    el: "#app",
    data:
    {
```

```
            tableData: [{
                ID: 1,
                name: "张三",
                grade: 80,
                phone: "12345678901",
                adress: "武汉市洪山区某某街道"
            }, {
                ID: 2,
                name: "李四",
                grade: 90,
                phone: "12345678901",
                adress: "武汉市洪山区某某街道"
            }, {
                ID: 3,
                name: "王五",
                grade: 85,
                phone: "12345678901",
                adress: "武汉市洪山区某某街道"
            }, {
                ID: 4,
                name: "赵六",
                grade: 95,
                phone: "12345678901",
                adress: "武汉市洪山区某某街道"
            }, {
                ID: 5,
                name: "周七",
                grade: 85,
                phone: "12345678901",
                adress: "武汉市洪山区某某街道"
            }, {
                ID: 6,
                name: "吴八",
                grade: 90,
                phone: "12345678901",
                adress: "武汉市洪山区某某街道"
            }],
            multipleSelection: []
        },
        methods:{
            //单击表格中多选框时的响应事件，val 为一个数组，存放当前所选的多行数据
            handleSelectionChange:function(val){
                multipleSelection=val;
                console.log(multipleSelection);
            }
        }
    })
```

多选样式对应的标签代码如下：

```
<el-table :data="tableData" height="250"
    @selection-change="handleSelectionChange">
    <el-table-column type="selection" width="50">
    </el-table-column>
    <el-table-column prop="ID" label="ID" width="60" fixed="left">
    </el-table-column>
    <el-table-column prop="name" label="姓名" width="60">
    </el-table-column>
    <el-table-column prop="grade" label="成绩" width="60">
    </el-table-column>
    <el-table-column prop="phone" label="电话" width="120">
    </el-table-column>
    <el-table-column prop="adress" label="家庭住址" width="160">
    </el-table-column>
</el-table>
```

（12）排序。在使用表格的排序功能时，只需要在需要排序的列标签中添加 sortable 属性即可。通过设置 el-table 标签中的 default-sort 属性可以指定排序的列和排序规则，如果 default-sort 属性中 order 的取值为 descending，则表示降序；如果 default-sort 属性中 order 的取值为 ascending，则表示升序（默认为升序）。

本示例首先创建 Vue，该 Vue 实例与固定列和表头的 Vue 实例一样，故不再展示 Vue 实例的代码；然后结合地图展示排序的样式，如图 4-13 所示。

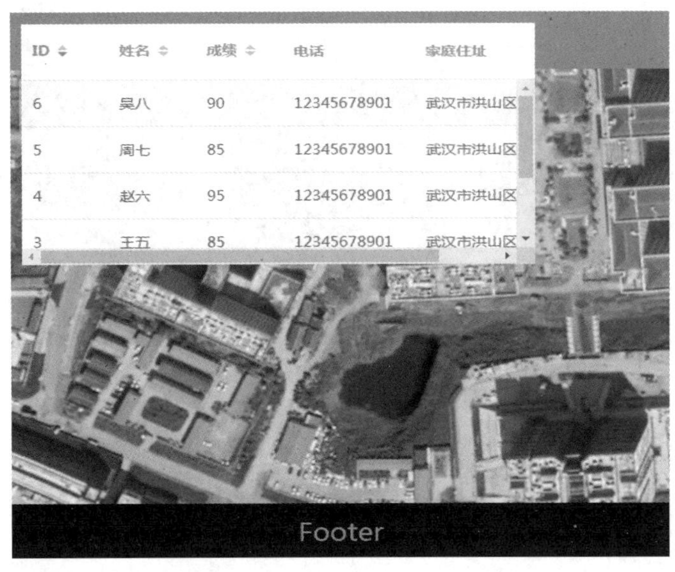

图 4-13　排序的样式

排序样式对应的标签代码如下：

```
<el-table :data="tableData" height="250"
:default-sort = "{prop: 'ID', order: 'descending'}">
```

```
        <el-table-column prop="ID" label="ID" width="80" sortable>
        </el-table-column>
        <el-table-column prop="name" label="姓名" width="80" sortable>
        </el-table-column>
        <el-table-column prop="grade" label="成绩" width="80" sortable>
        </el-table-column>
        <el-table-column prop="phone" label="电话" width="120">
        </el-table-column>
        <el-table-column prop="adress" label="家庭住址" width="160">
        </el-table-column>
    </el-table>
```

（13）筛选。在使用表格的筛选功能时，需要首先在某列的标签中设置 filter 属性并赋值一个对象数组，然后通过 filter-method 属性来定义一个筛选方法。

本示例首先创建 Vue 实例，data 中的 tableData 是一个对象数组，每个对象都表示表格中的一行数据，methods 中的 filterHandler 是表格筛选的响应事件；然后结合地图展示筛选的样式，如图 4-14 所示。代码如下：

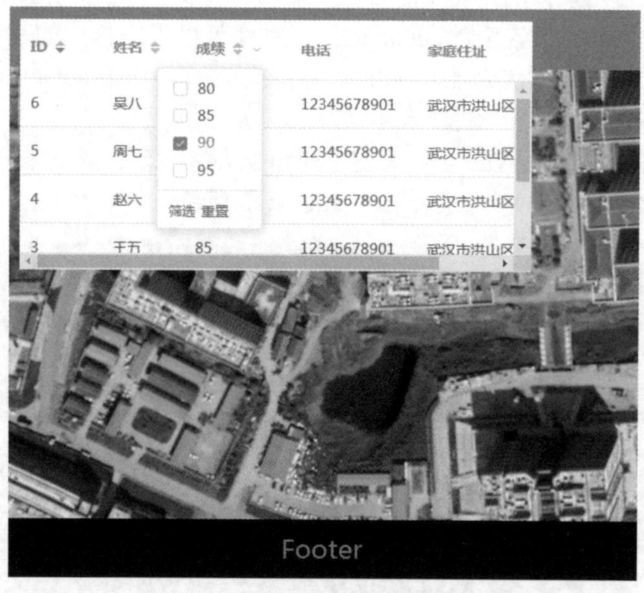

图 4-14　筛选的样式

```
    var vm = new Vue({
        el: "#app",
        data:
        {
            tableData: [{
                ID: 1,
                name: "张三",
                grade: 80,
                phone: "12345678901",
                adress: "武汉市洪山区某某街道"
```

```
        },{
            ID: 2,
            name: "李四",
            grade: 90,
            phone: "12345678901",
            adress: "武汉市洪山区某某街道"
        },{
            ID: 3,
            name: "王五",
            grade: 85,
            phone: "12345678901",
            adress: "武汉市洪山区某某街道"
        },{
            ID: 4,
            name: "赵六",
            grade: 95,
            phone: "12345678901",
            adress: "武汉市洪山区某某街道"
        },{
            ID: 5,
            name: "周七",
            grade: 85,
            phone: "12345678901",
            adress: "武汉市洪山区某某街道"
        },{
            ID: 6,
            name: "吴八",
            grade: 90,
            phone: "12345678901",
            adress: "武汉市洪山区某某街道"
        }]

    },
    methods: {
        //表格筛选的响应事件
        filterHandler:function(value, row, column) {
            var property = column['property'];
            return row[property] === value;
        }
    }
})
```

筛选样式对应的标签代码如下：

```
<el-table :data="tableData" height="250" :default-sort="{prop: 'ID', order: 'descending'}">
    <el-table-column prop="ID" label="ID" width="80" sortable>
    </el-table-column>
```

```
        <el-table-column prop="name" label="姓名" width="80" sortable>
        </el-table-column>
        <el-table-column prop="grade" label="成绩" width="100" sortable
            :filters="[{text:80,value:80},{text:85,value:85},{text:90,value:90},{text:95,value:95}]"
            :filter-method="filterHandler">
        </el-table-column>
        <el-table-column prop="phone" label="电话" width="120">
        </el-table-column>
        <el-table-column prop="adress" label="家庭住址" width="160">
        </el-table-column>
    </el-table>
```

（14）自定义列模板。通过自定义列模板可以自定义列的显示内容，自定义列模板可以和其他组件组合使用。本示例首先隐藏住址这一列的数据，然后在鼠标光标移动到姓名处时，会弹出该姓名及其对应住址的对话框。本示例是通过在姓名这一列的标签中嵌入一个模板来实现的。

本示例首先创建 Vue，该 Vue 实例与固定列和表头的 Vue 实例一样，故不再展示 Vue 实例的代码；然后结合地图展示自定义列模板的样式，如图 4-15 所示。代码如下：

图 4-15　自定义列模板的样式

```
    <el-table :data="tableData" height="250">
        <el-table-column prop="ID" label="ID" width="60" fixed="left">
        </el-table-column>
        <el-table-column prop="name" label="姓名" width="60">
            <!-- slot-scope 用于获取表格的内部数据 -->
            <template slot-scope="scope">
                <el-popover trigger="hover" placement="top">
                    <p>姓名: {{ scope.row.name }}</p>
                    <p>住址: {{ scope.row.address }}</p>
```

```
                <div slot="reference">
                    <el-tag size="medium">{{ scope.row.name }}</el-tag>
                </div>
            </el-popover>
        </template>
    </el-table-column>
    <el-table-column prop="grade" label="成绩" width="60">
    </el-table-column>
    <el-table-column prop="phone" label="电话" width="120">
    </el-table-column>
</el-table>
```

（15）展开行。展开行是指当表格的横向内容过多但又不想使用滚动条时，我们可以在表格中显示主要的内容，然后通过展开行来显示该行的全部信息。要实现此功能，首先需要在el-table-column标签中将type属性设置为expand，然后在该标签中嵌入模板并设置其属性slot-scope，紧接着将需要展开的行内容放在el-form标签中即可。

本示例首先创建Vue，该Vue实例与固定列和表头的Vue实例一样，故不再展示Vue实例的代码；然后结合地图展示展开行的样式，如图4-16所示。

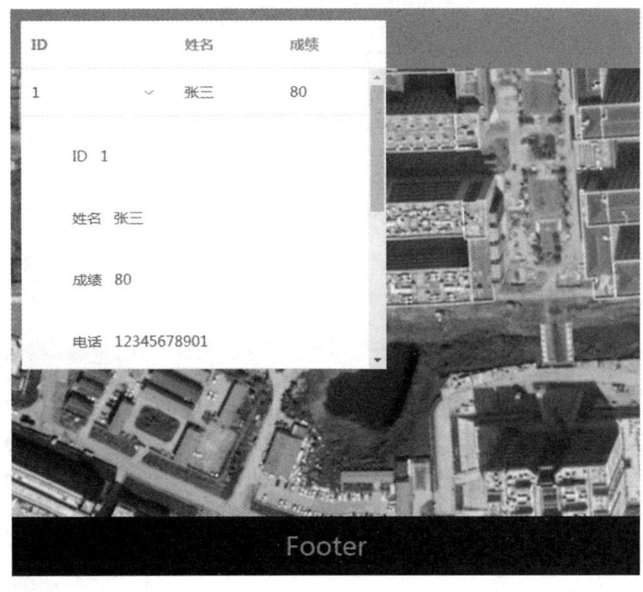

图4-16　展开行的样式

展开行样式对应的标签代码如下：

```
<el-table :data="tableData" height="350">
    <el-table-column type="expand">
        <!-- slot-scope 用于获取表格的内部数据 -->
        <template slot-scope="props">
            <el-form label-position="left">
                <el-form-item label="ID">
                    <span>{{ props.row.ID }}</span>
```

```
                </el-form-item>
                <el-form-item label="姓名">
                    <span>{{ props.row.name }}</span>
                </el-form-item>
                <el-form-item label="成绩">
                    <span>{{ props.row.grade }}</span>
                </el-form-item>
                <el-form-item label="电话">
                    <span>{{ props.row.phone }}</span>
                </el-form-item>
                <el-form-item label="住址">
                    <span>{{ props.row.address }}</span>
                </el-form-item>
            </el-form>
        </template>
    </el-table-column>
    <el-table-column prop="ID" label="ID" width="100" fixed="left">
    </el-table-column>
    <el-table-column prop="name" label="姓名" width="100">
    </el-table-column>
    <el-table-column prop="grade" label="成绩" width="100">
    </el-table-column>
</el-table>
```

（16）树形数据与懒加载。当表格中的行包含 children 字段时，该行的数据可看成树形数据。当渲染树形数据时，必须指定 row-key。

本示例首先创建 Vue 实例，data 中的 tableData 是一个对象数组，每个对象都表示表格中的一行数据；然后结合地图展示树形数据与懒加载的样式，如图 4-17 所示。代码如下：

图 4-17　树形数据与懒加载的样式

```
var vm = new Vue({
    el: "#app",
    data:
    {
        tableData: [{
            ID: 1,
            name: "张三",
            grade: 80,
            phone: "12345678901",
            address: "武汉市洪山区某某街道"
        }, {
            ID: 2,
            name: "李四",
            grade: 90,
            phone: "12345678901",
            address: "武汉市洪山区某某街道",
            children: [
                {
                    ID: 2.1,
                    name: "李四 1",
                    grade: 90,
                    phone: "12345678901",
                    address: "武汉市洪山区某某街道",
                },
                {
                    ID: 2.2,
                    name: "李四 2",
                    grade: 90,
                    phone: "12345678901",
                    address: "武汉市洪山区某某街道",
                }]
        }, {
            ID: 3,
            name: "王五",
            grade: 85,
            phone: "12345678901",
            address: "武汉市洪山区某某街道"
        }, {
            ID: 4,
            name: "赵六",
            grade: 95,
            phone: "12345678901",
            address: "武汉市洪山区某某街道"
        }, {
            ID: 5,
            name: "周七",
            grade: 85,
```

```
                phone: "12345678901",
                address: "武汉市洪山区某某街道"
            }, {
                ID: 6,
                name: "吴八",
                grade: 90,
                phone: "12345678901",
                address: "武汉市洪山区某某街道"
            }],
        }
})
```

树形数据与懒加载样式对应的标签代码如下：

```
<el-table :data="tableData" height="350" row-key="ID">
    <el-table-column prop="ID" label="ID" width="60">
    </el-table-column>
    <el-table-column prop="name" label="姓名" width="60">
    </el-table-column>
    <el-table-column prop="grade" label="成绩" width="60">
    </el-table-column>
    <el-table-column prop="phone" label="电话" width="120">
    </el-table-column>
    <el-table-column prop="address" label="家庭住址" width="160">
    </el-table-column>
</el-table>
```

（17）自定义表头。自定义表头是指可以在表头上添加其他组件，如搜索框。在某一 el-table-column 标签中嵌入相关的模板即可实现自定义表头。需要注意的是，必须将模板的 slot 属性设置为 header。

本示例首先创建 Vue 实例，data 中的 tableData 是一个对象数组，每个对象都表示表格中的一行数据，search 是输入框的数据，methods 中的 handleEdit 是单击每一行“编辑”按钮时的响应事件，该事件接收两个参数，这两个参数是通过模板中的 slot-scope 属性来传递的，第一个参数是对应的行号，第二个参数是对应的行数据；然后结合地图展示自定义表头的样式，如图 4-18 所示。代码如下：

```
var vm = new Vue({
    el: "#app",
    data:
    {
        search: "",
        tableData: [{
            ID: 1,
            name: "张三",
            grade: 80,
            phone: "12345678901",
        }, {
```

```
                ID: 2,
                name: "李四",
                grade: 90,
                phone: "12345678901",
            }, {
                ID: 3,
                name: "王五",
                grade: 85,
                phone: "12345678901"
            }, {
                ID: 4,
                name: "赵六",
                grade: 95,
                phone: "12345678901"
            }],
        },
        methods: {
            handleEdit: function (index, row) {
                console.log(index, row);
            }
        }
    })
```

图 4-18　自定义表头的样式

自定义表头样式对应的标签代码如下：

```
<el-table height="300"
    :data="tableData.filter(data => !search || data.name.toLowerCase().includes(search.toLowerCase()))">
    <el-table-column prop="ID" label="ID" width="60">
    </el-table-column>
```

```
        <el-table-column prop="name" label="姓名" width="60">
        </el-table-column>
        <el-table-column prop="grade" label="成绩" width="60">
        </el-table-column>
        <el-table-column prop="phone" label="电话" width="120">
        </el-table-column>
        <el-table-column align="right">
            <template slot="header" slot-scope="scope">
                <el-input v-model="search" size="mini" placeholder="输入关键字搜索" />
            </template>
            <template slot-scope="scope">
                <el-button type="primary" size="mini" @click="handleEdit(scope.$index, scope.row)">编辑
                </el-button>
            </template>
        </el-table-column>
    </el-table>
```

（18）表尾合计行。表尾合计行是指，如果表格中展示的是各类数字，则可以在表格的最后一行统计各列数字之和。只需要在 el-table 标签中添加 show-summary 属性即可实现表尾合计行功能。在默认情况下，对于合计行，第一列不进行数据求合操作，而是显示“合计”二字（可通过 sum-text 属性进行配置），其余列会对本列所有的数值进行求和操作，并显示出来。当然，用户也可以定义自己的合计逻辑。使用 summary-method 属性可以传入一个方法，并返回一个数组，这个数组中的各项会显示在合计行的各列中。

本示例首先创建 Vue 实例，data 中的 tableData 是一个对象数组，每个对象都表示表格中的一行数据；然后结合地图展示表尾合计行的样式，如图 4-19 所示。代码如下：

图 4-19　表尾合计行的样式

```
var vm = new Vue({
    el: "#app",
    data:
    {
```

```
            search: "",
            tableData: [{
                ID: 1,
                name: "张三",
                math: 80,
                english: 80,
                chinese: 92
            }, {
                ID: 2,
                name: "李四",
                math: 90,
                english: 90,
                chinese: 72
            }, {
                ID: 3,
                name: "王五",
                math: 85,
                english: 85,
                chinese: 84
            }, {
                ID: 4,
                name: "赵六",
                math: 95,
                english: 80,
                chinese: 93
            }],
        }
    })
```

表尾合计行样式对应的标签代码如下：

```
<el-table height="300" :data="tableData" show-summary sum-text="总分">
    <el-table-column prop="ID" label="ID" width="60">
    </el-table-column>
    <el-table-column prop="name" label="姓名" width="60">
    </el-table-column>
    <el-table-column prop="math" label="数学" width="60">
    </el-table-column>
    <el-table-column prop="english" label="英语" width="60">
    </el-table-column>
    <el-table-column prop="chinese" label="语文" width="60">
    </el-table-column>
</el-table>
```

（19）合并行或列。当多行或多列共用一个数据时，可以合并行或列。在 el-table 标签中设置 span-method 方法可合并行或列。span-method 方法的参数是一个对象，该对象包含了当前行、当前列、当前行号、当前列号四个属性。该方法既可以返回一个包含两个元素的数组，第一个元素表示 rowspan，第二个元素表示 colspan；也可以返回一个键名为 rowspan 和 colspan 的对象。

本示例首先创建 Vue 实例，data 中的 tableData 是一个对象数组，每个对象都表示表格中的一行数据，methods 中的 objectSpanMethod 是合并行或列的事件；然后结合地图展示合并行或列的样式，如图 4-20 所示。代码如下：

图 4-20　合并行或列的样式

```
var vm = new Vue({
    el: "#app",
    data:
    {
        search: "",
        tableData: [{
            ID: 1,
            name: "张三",
            math: 80,
            english: 80,
            chinese: 92
        }, {
            ID: 1,
            name: "李四",
            math: 90,
            english: 90,
            chinese: 72
        }, {
            ID: 2,
            name: "王五",
            math: 85,
            english: 85,
            chinese: 84
```

```
            }, {
                ID: 2,
                name: "赵六",
                math: 95,
                english: 80,
                chinese: 93
            }],
        },
        methods: {
            //行合并或列的事件
            objectSpanMethod: function ({ row, column, rowIndex, columnIndex }) {
                if (columnIndex === 0) {
                    if (rowIndex % 2 === 0) {
                        return {
                            rowspan: 2,
                            colspan: 1
                        };
                    }
                    else {
                        return {
                            rowspan: 0,
                            colspan: 0
                        };
                    }
                }
            },
        }
    })
```

合并行或列样式对应的标签代码如下：

```
<el-table height="300" :data="tableData" :span-method="objectSpanMethod">
    <el-table-column prop="ID" label="ID" width="60">
    </el-table-column>
    <el-table-column prop="name" label="姓名" width="60">
    </el-table-column>
    <el-table-column prop="math" label="数学" width="60">
    </el-table-column>
    <el-table-column prop="english" label="英语" width="60">
    </el-table-column>
    <el-table-column prop="chinese" label="语文" width="60">
    </el-table-column>
</el-table>
```

（20）自定义索引。根据前面的介绍，我们已经知道，将el-table标签中第一个el-table-column标签中的type属性设置为index，可以获取表格的索引号。如果要自定义索引，则只需要在该标签中设置index属性，该属性对应了一个方法，在该方法中对索引号进行自定义。

本示例首先创建 Vue 实例，data 中的 tableData 是一个对象数组，每个对象都表示表格中的一行数据，methods 中的 indexMethod 是自定义索引事件，该方法的参数为索引号；然后结合地图展示自定义索引的样式，如图 4-21 所示。代码如下：

图 4-21　自定义索引的样式

```
var vm = new Vue({
    el: "#app",
    data:
    {
        tableData: [{
            ID: 1,
            name: "张三",
            grade: 80,
            phone: "12345678901",
            address: "武汉市洪山区某某街道"
        }, {
            ID: 2,
            name: "李四",
            grade: 90,
            phone: "12345678901",
            address: "武汉市洪山区某某街道"
        }, {
            ID: 3,
            name: "王五",
            grade: 85,
            phone: "12345678901",
            address: "武汉市洪山区某某街道"
        }, {
            ID: 4,
```

```
            name: "赵六",
            grade: 95,
            phone: "12345678901",
            address: "武汉市洪山区某某街道"
        }]
    },
    methods: {
        //自定义索引事件
        indexMethod:function(index){
            return index*2;
        }
    }
})
```

自定义索引样式对应的标签代码如下：

```
<el-table :data="tableData" height="250">
    <el-table-column type="index" :index="indexMethod">
    </el-table-column>
    <el-table-column prop="ID" label="ID" width="60" fixed="left">
    </el-table-column>
    <el-table-column prop="name" label="姓名" width="60">
    </el-table-column>
    <el-table-column prop="grade" label="成绩" width="60">
    </el-table-column>
    <el-table-column prop="phone" label="电话" width="120">
    </el-table-column>
    <el-table-column prop="address" label="家庭住址" width="160">
    </el-table-column>
</el-table>
```

4.2 标签（Tag）

Element 为标签提供了不同的样式与用法，本节将从基础用法、可移除标签、动态编辑标签、不同尺寸、不同主题等方面讲解标签的样式与用法。本节首先结合地图展示标签的样式，如图 4-22 所示；然后对标签的样式与用法进行讲解。

本示例创建了 Vue 实例，其中 data 中的 inputVisible 用于切换动态编辑标签中的输入框与按钮的状态，inputValue 是动态编辑标签中输入框 v-model 属性绑定的变量，dynamicTags 是动态编辑标签需要渲染的标签数据，tags1 是其他标签样式需要渲染的标签数据，methods 中的 handleClose1 是可移除标签样式中的标签移除事件，handleClose2 是动态编辑标签样式中的标签移除事件，showInput 是动态编辑标签样式中的“新建标签”按钮的单击响应事件，handleInputConfirm 是按下回车键和鼠标焦点离开输入框时的响应事件，在可移除标签样式中，若移除“标签一（地图）”，则地图会消失，如图 4-23 所示。代码如下：

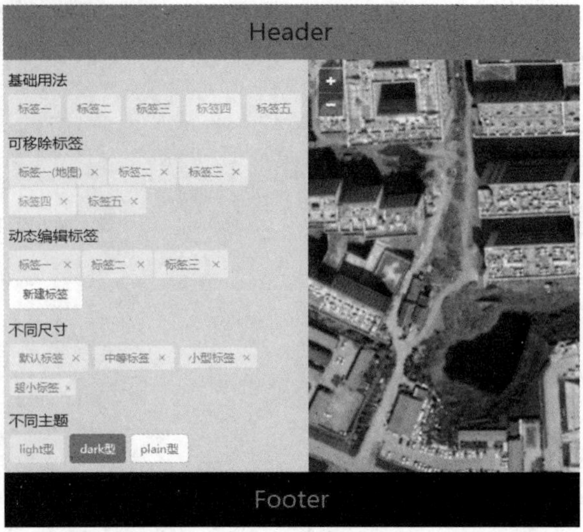

图 4-22　标签的样式

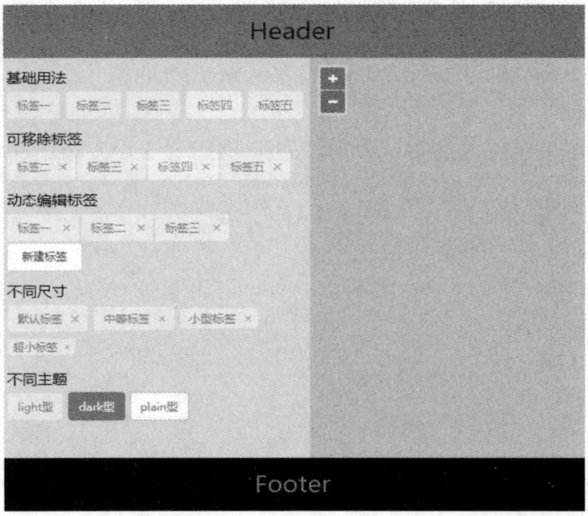

图 4-23　标签移除与地图消失

```
var vm = new Vue({
    el: "#app",
    data: {
        inputVisible: false,
        inputValue: '',
        tags1: [
            { name: '标签一（地图）', type: '' },
            { name: '标签二', type: 'success' },
            { name: '标签三', type: 'info' },
            { name: '标签四', type: 'warning' },
            { name: '标签五', type: 'danger' }
        ],
        dynamicTags: ['标签一', '标签二', '标签三'],
    },
```

```
    methods: {
        //可移除标签样式中的标签移除事件
        handleClose1: function (tag) {
            this.tags1.splice(this.tags1.indexOf(tag), 1);
            if(tag.name=="标签一（地图）"){
                imageLayer.setVisible(false);
            }
        },
        //动态编辑标签样式中的标签移除事件
        handleClose2: function (tag) {
            this.dynamicTags.splice(this.dynamicTags.indexOf(tag), 1);
        },
        // "新建标签"按钮的单击响应事件
        showInput: function () {
            this.inputVisible = true;
        },
        //按下回车键和鼠标焦点离开输入框时的响应事件
        handleInputConfirm() {
            let inputValue = this.inputValue;
            if (inputValue) {
                this.dynamicTags.push(inputValue);
            }
            this.inputVisible = false;
            this.inputValue = '';
        }
    }
})
```

（1）基础用法。标签多用于标注和选择，Element 通过 el-tag 标签实现了标签，el-tag 标签中的 type 属性用于选择标签的类型，el-tag 标签中的 color 属性用于定义标签的背景颜色。代码如下：

```
<el-tag>标签一</el-tag>
<el-tag type="success">标签二</el-tag>
<el-tag type="info">标签三</el-tag>
<el-tag type="warning">标签四</el-tag>
<el-tag type="danger">标签五</el-tag>
```

（2）可移除标签。在 el-tag 标签中设置 closable 属性，根据待移除标签中的 close 事件即可移除该标签。在默认的情况下，移除标签时会附带渐变动画效果，如果不想使用渐变动画效果，则可通过设置 disable-transitions 属性来取消渐变动画效果。代码如下：

```
<el-tag v-for="tag of tags1" :type="tag.type" closable :disable-transitions="false"
    @close="handleClose1(tag)">{{tag.name}}
</el-tag>
```

（3）动态编辑标签。动态编辑标签是指通过一个按钮或输入框来实现标签的动态增加，通过可移除标签属性可以实现标签的动态移除效果。代码如下：

```
<el-tag v-for="tag of dynamicTags" closable :disable-transitions="false"
      @close="handleClose2(tag)">{{tag}}
</el-tag>
<!-- @keyup.enter.native 为按下回车键时的响应事件，@blur 为鼠标焦点离开输入框时的响应事件 -->
<el-input v-if="inputVisible" v-model="inputValue" size="small"
      @keyup.enter.native="handleInputConfirm" @blur="handleInputConfirm">
</el-input>
<el-button v-else size="small" @click="showInput">新建标签
</el-button>
```

（4）不同尺寸。Element 为标签提供了四种尺寸，分别为 large、medium、small、mini，默认的尺寸为 large。通过 el-tag 标签中的 size 属性可以设置标签的尺寸。代码如下：

```
<el-tag closable>默认标签</el-tag>
<el-tag size="medium" closable>中等标签</el-tag>
<el-tag size="small" closable>小型标签</el-tag>
<el-tag size="mini" closable>超小标签</el-tag>
```

（5）不同主题。Element 为标签提供了三种主题，分别为 dark、light、plain，默认的主题是 light。通过 el-tag 标签中的 effect 属性可以设置标签的主题。代码如下：

```
<el-tag effect="light">light 型</el-tag>
<el-tag effect="dark">dark 型</el-tag>
<el-tag effect="plain">plain 型</el-tag>
```

4.3 进度条（Progress）

Element 为进度条提供了不同的样式与用法，本节将从线性进度条、百分比内显、自定义颜色、环形进度条、仪表盘形进度条等方面讲解进度条的样式与用法。本节首先结合地图展示进度条的样式，如图 4-24 所示；然后对进度条的样式与用法进行讲解。

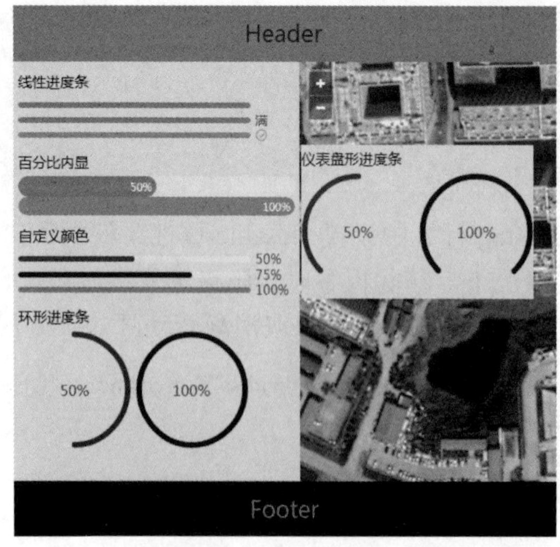

图 4-24　进度条样式

本示例创建了 Vue 实例，其中 data 中的 percentage 是线性进度条的 percentage 属性所对应的变量；methods 中的 format1 是线性进度条样式中第一个进度条的方法，用于动态减少进度条，format2 是线性进度条样式中第二个进度条的方法，用于指定进度条的文字。线性进度条的第一个进度条会自动从 100%减少到 0，在此过程中，地图的透明度也会从 100%减少到 0，如图 4-25 所示。代码如下：

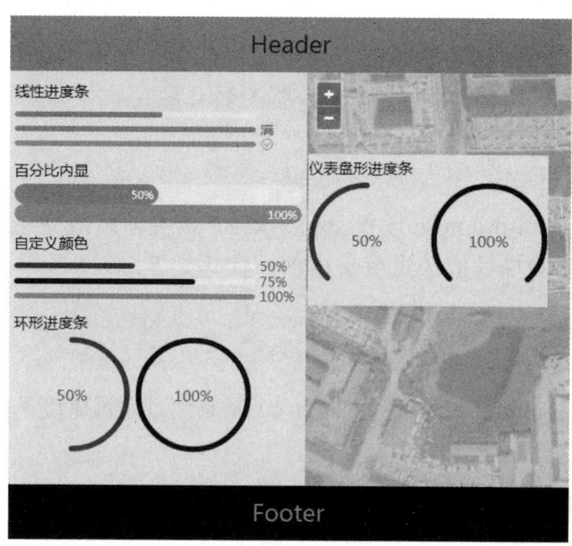

图 4-25　进度条与地图透明度

```
var vm = new Vue({
    el: "#app",
    data: {
        percentage: 100
    },
    methods: {
        //线性进度条样式中第一个进度条的方法
        format1: function () {
            //通过定时器来使进度条减少
            var timer = setInterval( ()=> {
                if (this.percentage >0) {
                    this.percentage -= 10;
                    //通过进度条的值和地图的透明度进行联动
                    imageLayer.setOpacity(this.percentage/100);
                }
                else {
                    clearInterval(timer)
                }
            }, 1000);
        },
        //线性进度条样式中第二个进度条的方法
        format2: function (percentage) {
```

```
                return percentage === 100 ? '满' : `${percentage}%`;
            }
        }
    })
```

（1）线性进度条。Element 通过对 el-progress 标签的封装提供了进度条，只需要在 el-progress 标签中设置 percentage 属性并进行赋值即可，赋值的范围为 0～100，status 用于控制进度条的当前状态。代码如下：

```
<el-progress :percentage="percentage" :format="format1"></el-progress>
<el-progress :percentage="100" :format="format2"></el-progress>
<el-progress :percentage="100" status="success"></el-progress>
```

（2）百分比内显。在 el-progress 设置 stroke-width 属性即可改变进度条的高度，并可通过 text-inside 属性将进度条的描述放在进度条内部。代码如下：

```
<el-progress :percentage="50" :stroke-width="22" text-inside></el-progress>
<el-progress :percentage="100" status="success" :stroke-width="20" text-inside></el-progress>
```

（3）自定义颜色。在 el-progress 标签中添加 color 属性可设置进度条的背景颜色。代码如下：

```
<el-progress :percentage="50" color="red"></el-progress>
<el-progress :percentage="75" color="rgb(0,0,255)"></el-progress>
<el-progress :percentage="100" color="#00ff00"></el-progress>
```

（4）环形进度条。将 el-progress 标签中的 type 属性设置为 circle，可以将进度条设置为环形进度条，并可以通过 width 属性来指定环形进度条的大小。代码如下：

```
<el-progress :percentage="50" color="red" type="circle"></el-progress>
<el-progress :percentage="100" color="rgb(0,0,255)" type="circle"></el-progress>
```

（5）仪表盘形进度条。将 el-progress 标签中的 type 属性设置为 dashboard，可以将进度条设置为仪表盘形进度条。代码如下：

```
<el-progress :percentage="50" color="red" type="dashboard"></el-progress>
<el-progress :percentage="100" color="rgb(0,0,255)" type="dashboard"></el-progress>
```

4.4 树形（Tree）组件

Element 为树形组件提供了不同的样式与用法，本节将从基础用法、可选择、默认展开和默认选中、禁用状态、自定义节点内容、节点过滤、可拖曳节点等方面讲解树形组件的样式和用法。

由于树形组件所占的物理空间较大，因此本节分别用一个页面布局对每个样式进行讲解。

（1）基础用法。Element 通过对 el-tree 标签的封装提供了树形组件。本示例首先创建 Vue 实例，data 中的 options 属性是树形组件中的数据结构；然后结合地图展示基础用法的样式。如图 4-26 所示。代码如下：

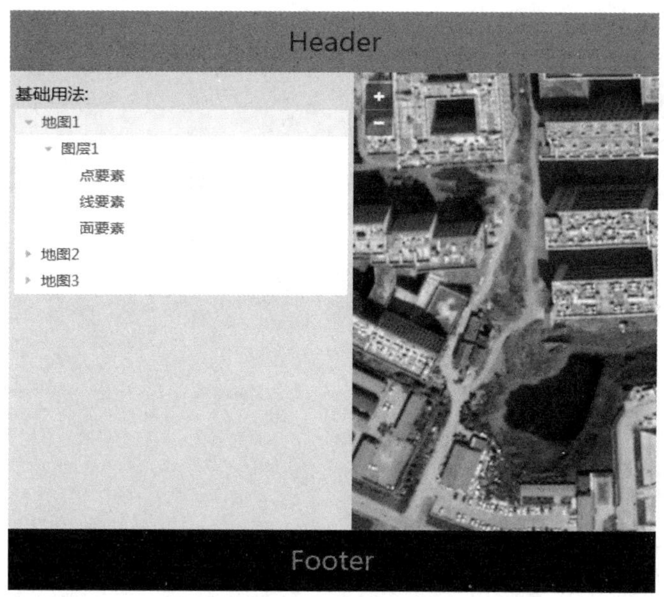

图 4-26　基础用法的样式

```
var vm = new Vue({
    el: "#app",
    data: {
        options: [{
            id: 1,
            label: '地图 1',
            children: [{
                id: 4,
                label: '图层 1',
                children: [{
                    id: 9,
                    label: '点要素',
                },
                {
                    id: 10,
                    label: '线要素',
                },
                {
                    id: 11,
                    label: '面要素',
                }]
            }]
        }, {
            id: 2,
            label: '地图 2',
            children: [{
                id: 5,
```

```
                    label: '图层 1',
                    children: [{
                        id: 12,
                        label: '点要素'
                    }]
                }, {
                    id: 6,
                    label: '图层 2',
                    children: [{
                        id: 13,
                        label: '点要素'
                    }]
                }]
            }, {
                id: 3,
                label: '地图 3',
                children: [{
                    id: 7,
                    label: '图层 1',
                    children: [{
                        id: 14,
                        label: '点要素'
                    }]
                }, {
                    id: 8,
                    label: '图层 2',
                    children: [{
                        id: 15,
                        label: '点要素'
                    }]
                }]
            }],
        }
})
```

基础用法样式对应的标签代码如下：

```
<el-tree :data="options"></el-tree>
```

（2）可选择。可选择样式适用于需要选择层级的场合，只需要在 el-tree 标签中添加 show-checkbox 属性即可实现可选择的样式，如图 4-27 所示。

本示例首先创建 Vue 实例，该示例中 data 的数据结构与基础用法中 data 的数据结构一样，这里不再展示 data 的代码；然后实现绘制图层（包括点的绘制图层、线的绘制图层和面的绘制图层）的控制，即根据是否勾选点要素、线要素和面要素前面的多选框来控制是否在绘制图层中显示点要素、线要素和面要素。图 4-28 在绘制图层中显示了点要素、线要素和面要素。

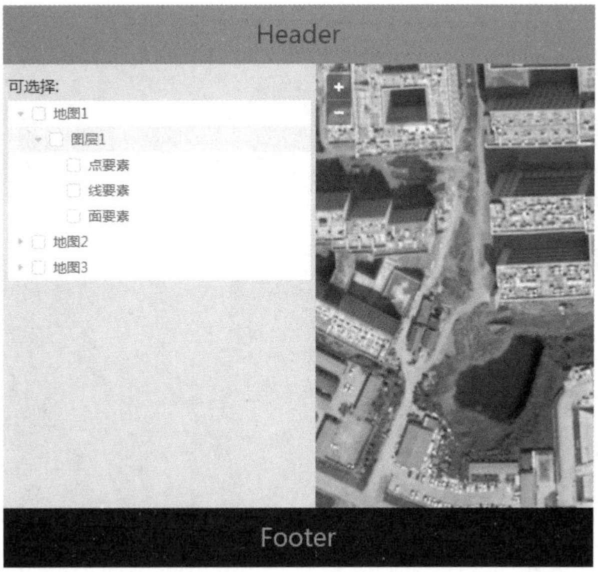

图 4-27　可选择的样式

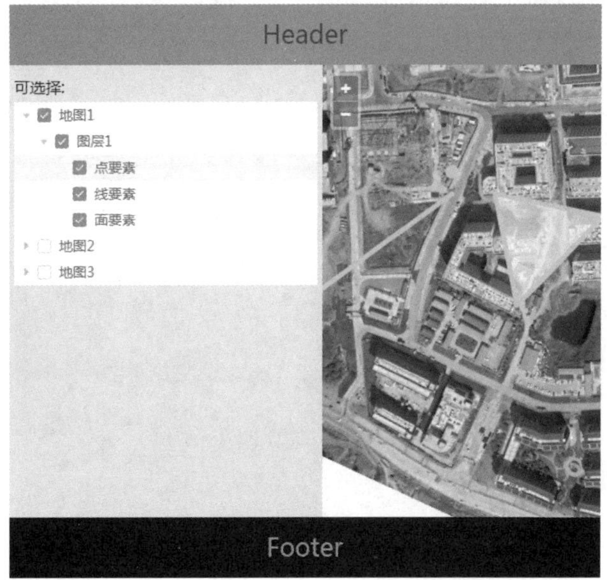

图 4-28　在绘制图层中显示点要素、线要素和面要素

创建绘制图层的代码如下：

```
//定义一个样式
var style = new ol.style.Style({
    //填充色
    fill: new ol.style.Fill({
        color: 'rgba(255, 255, 255, 0.7)'
    }),
    //边线颜色
```

```
        stroke: new ol.style.Stroke({
            color: '#ffcc33',
            width: 5
        }),
        //形状
        image: new ol.style.Circle({
            radius: 7,
            fill: new ol.style.Fill({
                color: '#ffcc33'
            })
        })
    });
    //定义点的绘制图层
    var PtSource = new ol.source.Vector();
    var PtVector = new ol.layer.Vector({
        source: PtSource
    });
    //定义线的绘制图层
    var lineSource = new ol.source.Vector();
    var lineVector = new ol.layer.Vector({
        source: lineSource
    })
    //定义面的绘制图层
    var polygonSource = new ol.source.Vector();
    var polygonVector = new ol.layer.Vector({
        source: polygonSource
    })

    //定义点
    var point = new ol.Feature({
        geometry: new ol.geom.Point([2000, 1900])
    });
    //给点添加样式
    point.setStyle(style);
    //将点的绘制图层添加到地图中
    map.addLayer(PtVector);
    //定义线
    var line = new ol.Feature({
        geometry: new ol.geom.LineString([[1300, 1300], [2000, 1800]])
    })
    //给线添加样式
    line.setStyle(style);
    //将线的绘制图层添加到地图中
    map.addLayer(lineVector);
    //定义面
    var polygon = new ol.Feature({
        geometry: new ol.geom.Polygon([[[2300, 1300], [2700, 1700], [2200, 1800]]])
```

```
});
//给面添加样式
polygon.setStyle(style);
//将面的绘制图层添加到地图中
map.addLayer(polygonVector)
```

获取和设置树形组件中的节点有两种方法：一种方法是通过 node 来获取和设置节点，另一种方法是通过 key 来获取和设置节点。如果要通过 key 来获取和设置节点，则必须在 el-tree 标签中设置 node-key。本示例通过 node 来获取和设置节点，代码如下：

```
methods: {
    //单击多选框时的响应事件
    handleCheckChange: function (data, checked) {
        //清空点、线、面的绘制图层
        PtSource.clear();
        lineSource.clear();
        polygonSource.clear();
        //通过树形组件的 getCheckedNodes 方法获取节点，然后进行遍历
        this.$refs.tree.getCheckedNodes().forEach(function (val, index) {
            switch (val.label) {
                case "点要素":
                    //在点的绘制图层中添加点要素
                    PtSource.addFeature(point);
                break;
                case "线要素":
                    //在线的绘制图层中添加线要素
                    lineSource.addFeature(line);
                break;
                case "面要素":
                    //在面的绘制图层中添加面要素
                    polygonSource.addFeature(polygon);
                break;
            }
        })
    }
}
```

可选择样式对应的标签代码如下：

```
<el-tree :data="options" node-key="id" ref="tree" show-checkbox
@check="handleCheckChange"></el-tree>
```

（3）默认展开和默认选中。在 el-tree 标签中设置 default-expanded-keys 和 default-checked-keys 属性可以实现默认展开和默认选择的样式。需要注意的是，此时必须设置 node-key。本示例首先创建 Vue 实例，该示例的 Vue 示例与基础用法的 Vue 实例一样，这里不再给出代码；然后结合地图展示默认展开和默认选中的样式，如图 4-29 所示。代码如下：

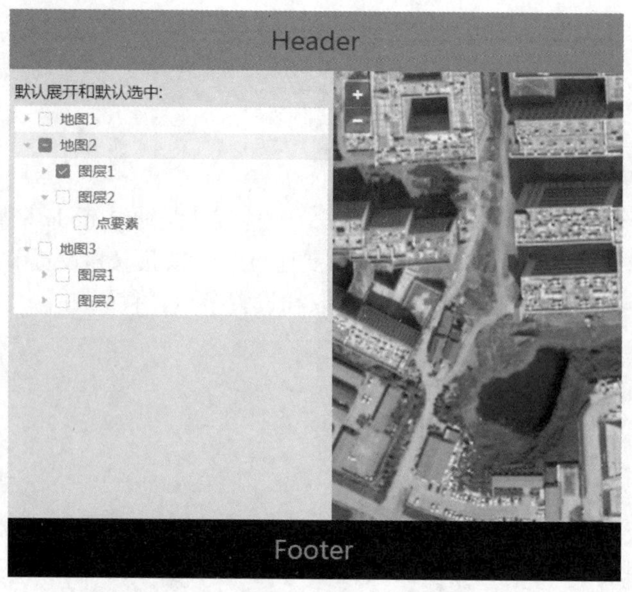

图 4-29　默认展开和默认选中的样式

```
<el-tree :data="options" node-key="id" show-checkbox :default-expanded-keys="[2,
3]" :default-checked-keys="[5]"></el-tree>
```

（4）禁用状态。在 el-tree 标签中添加 disabled 属性即可实现禁用状态的样式。本示例首先创建 Vue 实例，该示例的 Vue 示例与基础用法的 Vue 实例基本一样，唯一不同的是在 id 为 6 的节点中添加 disabled:true，这里不再给出具体代码；然后结合地图展示禁用状态的样式，如图 4-30 所示。

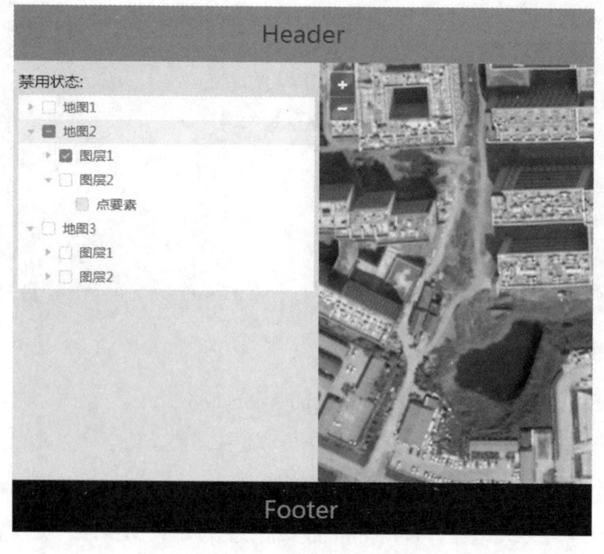

图 4-30　禁用状态的样式

（5）自定义节点内容。自定义节点内容是指用户可以增加或者删除树形组件中的节点，在节点区域添加按钮或者图标都可以控制节点的增加和删除。通过 render-content 和 scoped slot

这两种方法，可以实现自定义节点内容，在使用 scoped slot 方法时，会传入参数 node 和 data，这两个参数分别表示当前节点的对象和当前节点的数据。本示例使用的是 scoped slot 方法。

本示例首先创建 Vue 实例，该示例中 data 的数据结构与基础用法中 data 的数据结构一样，为了避免代码冗余，这里只给出了 methods 中的代码，其中 append 是单击“增加”按钮时的响应事件，remove 是单击“删除”按钮时的响应事件；然后结合地图展示自定义节点内容的样式，如图 4-31 所示。代码如下：

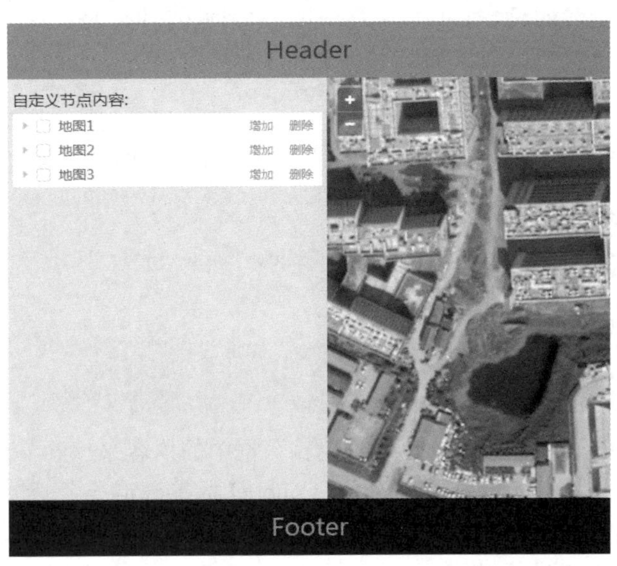

图 4-31　自定义节点内容的样式

```
methods: {
    //单击“增加”按钮时的响应事件
    append:function(data) {
        //创建一个子节点
        const newChild = { id: id++, label: 'test', children: [] };
        //如果该节点没有子节点，即该节点就是叶子节点，则给该节点创建一个子节点
        if (!data.children) {
            this.$set(data, 'children', []);
        }
        //将新建的子节点添加到该节点中
        data.children.push(newChild);
    },
    //单击“删除”按钮时的响应事件
    remove:function(node, data) {
        //获取将要删除节点的父节点
        var parent = node.parent;
        //获取将要删除节点的兄弟节点
        var children = parent.data.children || parent.data;
        //获取将要删除节点的索引
        var index = children.findIndex(function (value) {
            return value.id === data.id
```

```
            });
            //根据索引号删除对应的节点
            children.splice(index, 1);
        },
}
```

自定义节点内容样式对应的标签代码如下：

```
<el-tree :data="options" node-key="id" show-checkbox :expand-on-click-node="false">
    <span class="custom-tree-node" slot-scope="{ node, data }">
        <span>{{ node.label }}</span>
        <span>
            <el-button type="text" size="mini" @click="append(data)">
                增加
            </el-button>
            <el-button type="text" size="mini" @click="remove(node, data)">
                删除
            </el-button>
        </span>
</el-tree>
```

（6）节点过滤。通过节点过滤，可以根据输入框中的内容来搜索节点。要实现节点过滤，首先要添加一个输入框，即 el-input 标签，为 v-model 标签绑定一个变量，该变量通过 Vue 实例的 watch 方法对输入框中的内容进行实时监测；然后在 el-tree 标签中设置 filter-node-method 方法，通过该方法对节点进行过滤。

本示例首先创建 Vue 实例，将 data 中的 filterText 与 el-input 标签中的 v-model 属性进行联动，options 中的数据结构与基础用法的一样，这里只给出 methods 中的代码，其中 filterNode 是节点过滤方法，watch 是监测输入框中内容的方法；然后结合地图展示节点过滤的样式，如图 4-32 所示。代码如下：

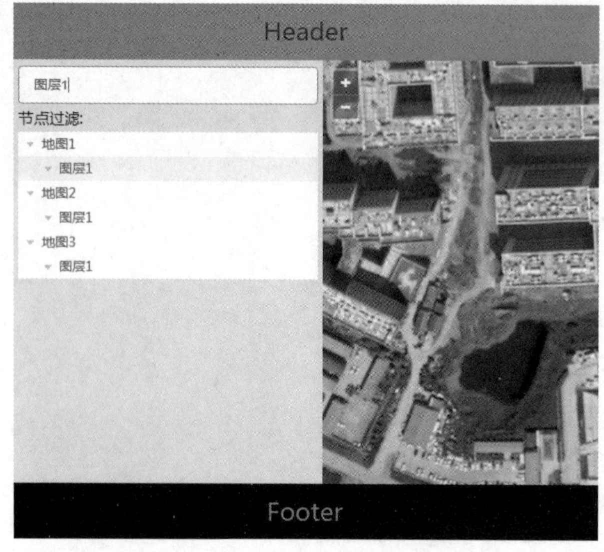

图 4-32　节点过滤的样式

```
methods: {
    //过滤节点的方法，value 为输入框输入的内容，data 为所有节点的数据
    filterNode:function(value, data) {
        //如果输入为空，则返回 true
        if (!value) {
            return true;
        }
        //若输入不为空，则在所有节点中查询输入框的内容
        else {
            return data.label.indexOf(value) !== -1;
        }
    }
},
//监测输入框数据的变动
watch: {
    filterText(val) {
        this.$refs.tree.filter(val);
    }
}
```

节点过滤样式对应的标签代码如下：

```
<el-input placeholder="输入关键字进行过滤" v-model="filterText">
</el-input>
<div>节点过滤:</div>
<el-tree :data="options" ref="tree" :filter-node-method="filterNode"></el-tree>
```

（7）可拖曳节点。可拖曳节点是指可以用鼠标拖曳的方式来移动节点的位置，在 el-tree 标签中添加 draggable 属性即可实现该功能。

本示例首先创建 Vue 实例，该示例的 Vue 示例与基础用法的 Vue 实例基本一样，这里不再给出代码；然后结合地图展示可拖曳节点的样式，如图 4-33 所示。

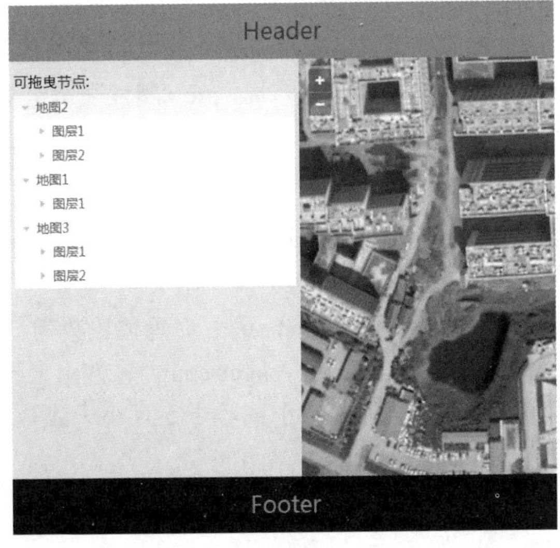

图 4-33　可拖曳节点的样式

可拖曳节点样式对应的标签代码如下：

```
<el-tree :data="options" draggable node-key="id"></el-tree>
```

4.5 分页（Pagination）

Element 为分页提供了不同的样式与用法，本节将从基础用法、设置最大页码按钮数、带有背景色的分页、附加功能、当只有一页时隐藏分页等方面讲解分页的样式与用法。本节首先结合地图展示分页的样式，如图 4-34 所示；然后对分页的样式与用法进行讲解。

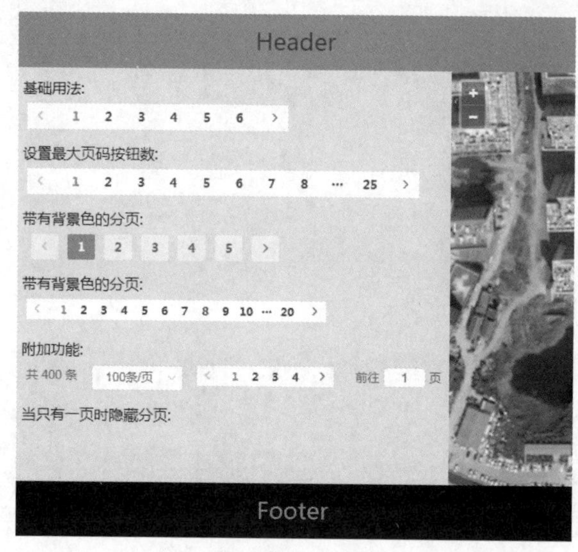

图 4-34　分页的样式

本示例首先创建 Vue 实例，methods 中的 handleChange 为基础用法中当前页码改变时的响应事件；然后通过对当前页码的控制，来实现地图视角的变化。

（1）基础用法。Element 通过对 el-pagination 标签的封装提供了分页组件，其中，el-pagination 标签中的 layout 属性用于组件的布局，prev 属性表示箭头向左翻页的按钮，pager 属性表示当前页码，next 属性表示箭头向右的按钮，page-size 属性表示每页显示的条目数，默认为 10，total 属性表示总的条目数，也就是说，页数等于 total 除以 page-size。代码如下：

```
<el-pagination layout="prev, pager, next" :total="60"
@current-change="handleChange"></el-pagination>
```

（2）设置最大页码按钮数。在默认情况下，最大页码按钮数为 7，只要超过 7 页，最后一个按钮与第 8 个按钮之间用省略号来表示。pager-count 属性用来设置最大页码按钮数，需要注意的是，pager-count 属性的取值只能是大于或等于 5 且小于或等于 21 的奇数。代码如下：

```
<el-pagination layout="prev, pager, next" :total="500" :pager-count="9"
    :page-size="20">
</el-pagination>
```

（3）带有背景色的分页。在 el-pagination 标签中添加 background 属性可以为分页设置背景色。代码如下：

```
<el-pagination layout="prev, pager, next" :total="500" :pager-count="9"
    :page-size="20">
</el-pagination>
```

（4）附加功能。附加功能是指通过在 el-pagination 标签中添加其他属性来实现更加复杂、完整的分页样式，例如可以在 layout 属性中添加 total、sizes、jumper 等。代码如下：

```
<el-pagination :page-sizes="[100, 200, 300, 400]"
    :page-size="100" layout="total, sizes, prev, pager, next, jumper" :total="400"
    :current-page="1" small>
</el-pagination>
```

（5）当只有一页时隐藏分页。当页码只有一页时，就有必要隐藏分页样式。在 el-pagination 标签中将 hide-on-single-page 属性设置为 true 即可实现此功能。代码如下：

```
<el-pagination :hide-on-single-page="true" :total="10" layout="prev, pager, next">
</el-pagination>
```

4.6　标注（Badge）

Element 为标注提供了不同的样式与用法，本节将从基础用法、最大值、自定义内容、小红点等方面讲解标注的样式与用法。本节首先结合地图展示标注的样式，如图 4-35 所示；然后对标注的样式与用法进行讲解。

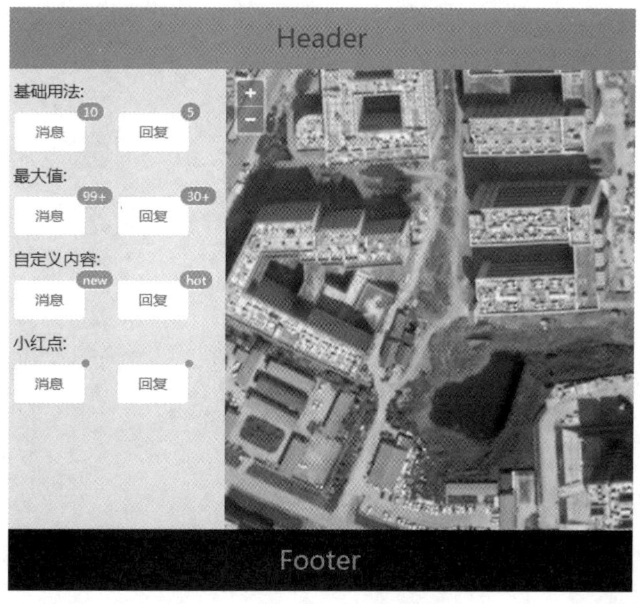

图 4-35　标注的样式

本示例首先创建 Vue 实例，data 中的 show 用于控制标注是否显示地图上的文本标注，vectorSource 是一个用于显示地图上文本标注的全局变量，methods 中的 handleClick 是基础用法中单击“消息”按钮时的响应事件；然后单击“消息”按钮，此时地图上会显示文本标注，并且“消息”按钮的标注也会消失，如图 4-36 所示。代码如下：

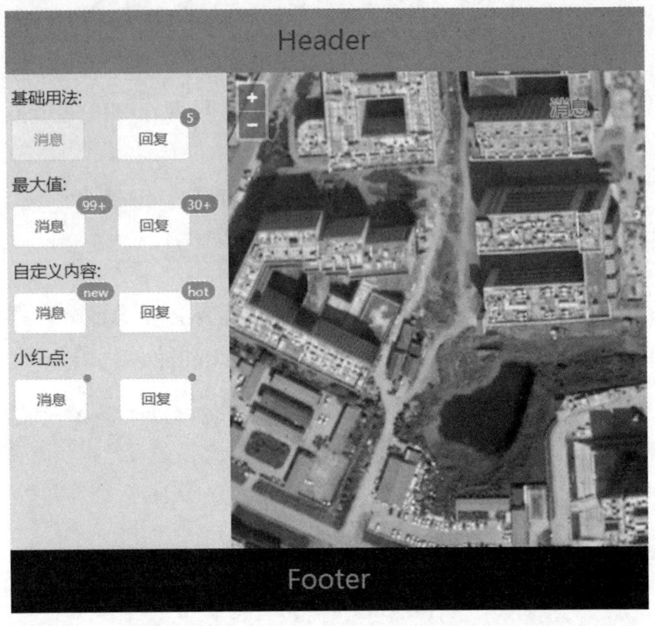

图 4-36　在地图上显示文本标注并且使“消息”按钮的标注消失

```
var vm = new Vue({
    el: "#app",
    data: {
        show:false,
        //矢量标注的数据源
        vectorSource: new ol.source.Vector()
    },
    methods: {
        handleClick: function () {
            this.show=true;
            //创建矢量标注样式函数
            var createLabelStyle = function (feature) {
                return new ol.style.Style({
                    text: new ol.style.Text({
                        //位置
                        textAlign: 'center',
                        //基准线
                        textBaseline: 'middle',
                        //文字样式
                        font: 'normal 18px 微软雅黑',
                        //文本内容
```

```
                    text: feature.get('name'),
                    //文本填充样式（即文字颜色）
                    fill: new ol.style.Fill({ color: '#aa4400' }),
                    stroke: new ol.style.Stroke({ color: '#ffcc33', width: 2 })
                })
            });
        }
        //创建点的坐标
        var pt = [2750, 1900];
        //实例化 Vector 要素，通过矢量图层添加到地图容器中
        var iconFeature = new ol.Feature({
            geometry: new ol.geom.Point(pt),
            name: "消息"
        })
        iconFeature.setStyle(createLabelStyle(iconFeature));
        //添加文本标注之前先清空原来的标注
        this.vectorSource.clear();
        this.vectorSource.addFeature(iconFeature);
        //矢量标注图层
        var vectorLayer = new ol.layer.Vector({
            source: this.vectorSource
        });
        map.addLayer(vectorLayer);
    }
  }
})
```

（1）基础用法。Element 通过对 el-badge 标签的封装提供了标注组件，该标签中的 value 属性是标注的内容，标注的类型可以是 Number，也可以是 String。代码如下：

```
<el-badge :value="10" class="item">
    <el-button>消息</el-button>
</el-badge>
<el-badge :value="5" type="primary" class="item" :hidden="show">
    <el-button>回复</el-button>
</el-badge>
```

（2）最大值。在 el-badge 标签中设置 max 属性可以定义标注的最大值，当 value 的对应值大于标注的最大值时，会用加号来表示。需要注意的是，只有当 value 的对应值类型为 Number 时，标注的最大值才会生效。代码如下：

```
<el-badge :value="200" class="item" :max="99">
    <el-button>消息</el-button>
</el-badge>
<el-badge :value="50" type="primary" class="item" :max="30">
    <el-button>回复</el-button>
</el-badge>
```

（3）自定义内容。用户可以为 value 属性对应的值赋予 String 类型的内容，如 new 和 hot（这也是常见的标注）。代码如下：

```
<el-badge value="new" class="item" :max="99">
    <el-button>消息</el-button>
</el-badge>
<el-badge value="hot" class="item" :max="30">
    <el-button size="评论">回复</el-button>
</el-badge>
```

（4）小红点。在 el-badge 标签中添加 is-dot 属性可以为标注的样式设置小红点。代码如下：

```
<el-badge is-dot class="item" :max="99">
    <el-button>消息</el-button>
</el-badge>
<el-badge is-dot class="item" :max="30">
    <el-button size="评论">回复</el-button>
</el-badge>
```

4.7 头像（Avatar）

Element 为头像提供了不同的样式与用法，本节将从基础用法、展示类型、图片加载失败的 fallback 行为、图片如何适应容器框等方面讲解头像的样式与用法。本节首先结合地图展示头像的样式，如图 4-37 所示；然后对头像的样式与用法进行讲解。

图 4-37　头像样式

本示例创建了 Vue 实例，methods 中的 errorHandler 是图片加载失败时的响应事件。代码如下：

```
var vm = new Vue({
    el: "#app",
    data: {
    },
    methods:{
        //图片加载失败时的响应事件
        errorHandler:function(){
            return true;
        }
    }
})
```

（1）基础用法。Element 通过 el-avatar 标签提供了头像组件。el-avatar 标签中的 src 属性表示图片的保存路径；shape 属性用于控制头像的形状，其取值为 circle 和 square，默认为 circle；size 属性用来控制头像的大小，既可以设置为具体的数值（Number 类型的数值），也可以通过 large、medium 和 small 等关键字来设置，默认为 large。需要注意的是，若通过数值来控制头像的大小，size 前需要加上冒号，否则在 Vue 实例中会将其看成 String 类型。代码如下：

```
<el-avatar :size="50" src=".\avatar.jpg"></el-avatar>
<el-avatar size="large" src=".\avatar.jpg"></el-avatar>
<el-avatar size="medium" src=".\avatar.jpg"></el-avatar>
<el-avatar size="small" src=".\avatar.jpg"></el-avatar>
<el-avatar :size="50" src=".\avatar.jpg" shape="square"></el-avatar>
<el-avatar size="large" src=".\avatar.jpg" shape="square"></el-avatar>
<el-avatar size="medium" src=".\avatar.jpg" shape="square"></el-avatar>
<el-avatar size="small" src=".\avatar.jpg" shape="square"></el-avatar>
```

（2）展示类型。Element 除了提供图片类型的头像，还提供了图标类型和文字类型的头像。图片类型是通过 src 属性设置的，图标类型是通过 icon 属性设置的，文字类型只需要在标签中写入需要的文字即可。

```
<el-avatar icon="el-icon-user-solid"></el-avatar>
<el-avatar src=".\avatar.jpg"></el-avatar>
<el-avatar>用户</el-avatar>
```

（3）图片加载失败的 fallback 行为。当图片加载失败时，Element 提供了补救的方法，这时首先需要在 el-avatar 标签中嵌入一个 img 标签，img 标签是图片加载失败时显示的图片；然后在 el-avatar 标签中添加 error 方法，该方法若返回 true 则表示该方法有效，若返回 false 则表示该方法无效。代码如下：

```
<el-avatar>
    <el-avatar :size="40" src="https://empty" @error="errorHandler">
        <img src="./fallbackImg.jpg" />
    </el-avatar>
</el-avatar>
```

（4）图片如何适应容器框。当头像是图片时，可以在 el-avatar 标签中设置 fit 属性来决定图片占用容器框的方式，fit 属性的取值可以是 fill、contain、cover、none 和 scale-down，默认的取值是 cover。其中，fill 表示图片拉伸充满整个容器框；contain 表示保持原有尺寸比例，长度和高度中较短的那条边和容器的大小一致，较长的那条边等比例缩放，可能会留有空白；cover 表示保持原有尺寸比例，宽度和高度中较长的那条边和容器的大小一致，较短的那条边等比例缩放，可能留有空白；none 表示保持原有尺寸比例；scale-down 表示保持原有尺寸比例，如果容器尺寸大于图片尺寸，则保持图片的原有尺寸，不会放大失真；如果容器尺寸小于图片尺寸，则用法跟 contain 一样。代码如下：

```
<el-avatar shape="square" size="medium" fit="fill" src=".\avatar.jpg">
</el-avatar>
<el-avatar shape="square" size="medium" fit="contain" src=".\avatar.jpg">
</el-avatar>
<el-avatar shape="square" size="medium" fit="cover" src=".\avatar.jpg">
</el-avatar>
<el-avatar shape="square" size="medium" fit="none" src=".\avatar.jpg">
</el-avatar>
<el-avatar shape="square" size="medium" fit="scale-down" src=".\avatar.jpg">
</el-avatar>
```

4.8 思考与练习题

（1）参考 4.1 节、4.4 节和 4.5 节，分别编写管理多个学生的表格和树形组件，将树形组件与分页关联起来，实现单击一个页码弹出一个新组件。

（2）参考 4.3 节、4.6 节和 4.7 节，实现单击“标注”按钮上传图片的功能，其中图片上传过程设置为 3 s，并将该过程与进度条关联起来。

第 5 章 通知组件

Element 为通知组件提供了非常丰富、详细的样式，本章将结合 Vue 和 OpenLayers 对 Element 的各个通知组件进行详细的介绍。

5.1 警告（Alert）

Element 为警告提供了不同的样式与用法，本节将从基础用法、主题、自定义关闭按钮、带有 icon、文字居中、带有辅助性文字介绍、带有 icon 和辅助性文字介绍等方面讲解警告的样式与用法。本节首先结合地图展示警告的样式，如图 5-1 所示；然后对警告的样式与用法进行讲解。

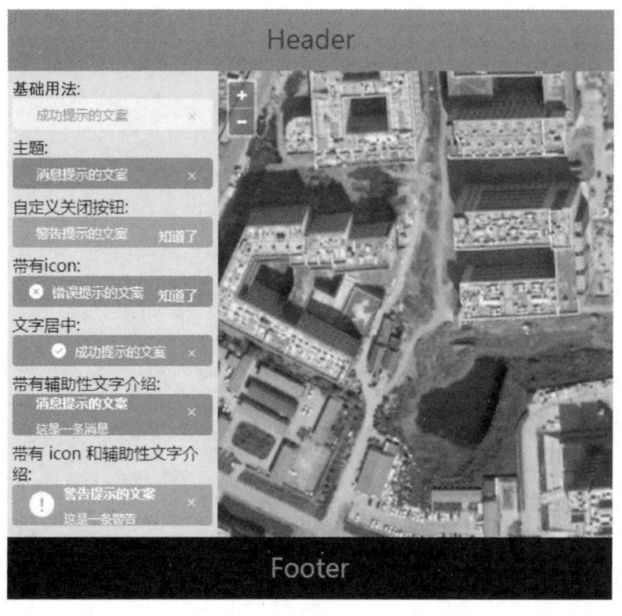

图 5-1　警告的样式

本示例首先创建 Vue 实例，methods 中的 handleClose 是基础用法中警告关闭的响应事件；

然后结合 OpenLayers 实现地图的隐藏操作，当关闭警告时，地图也随之消失，如图 5-2 所示所示。代码如下：

图 5-2　地图隐藏（一）

```
var vm = new Vue({
    el: "#app",
    data: {
    },
    methods: {
        handleClose: function () {
            //imageLayer 图层隐藏
            imageLayer.setVisible(false);
        }
    }
})
```

（1）基础用法。Element 通过封装 el-alert 标签提供了警告组件，el-alert 标签中的 title 属性是该 el-alert 标签的标题，type 属性决定了警告的类型，默认为 info。代码如下：

```
<el-alert title="成功提示的文案" type="success" @close="handleClose"></el-alert>
```

（2）主题。Element 为警告组件提供了两个不同的主题，在 el-alert 标签中设置 effect 属性可以选择主题，effect 属性的取值分别是 dark 和 light，默认的取值是 light。代码如下：

```
<el-alert title="消息提示的文案" type="info" effect="dark"></el-alert>
```

（3）自定义关闭按钮。将 el-alert 标签中的 closeable 属性设置为 false 可以让警告变得不可关闭，也可通过 close-text 标签用文本形式代替关闭图标。代码如下：

```
<el-alert title="警告提示的文案" close-text="知道了" type="warning" effect="dark"></el-alert>
```

（4）带有 icon。在 el-alert 标签中添加 show-icon 属性可以显示警告的图标，可以根据警告的 type 属性为该图标赋予不同的样式。代码如下：

```
<el-alert title="错误提示的文案" close-text="知道了" type="error" show-icon effect="dark"></el-alert>
```

（5）文字居中。在 el-alert 中添加 center 属性可以使警告的标题居中显示。代码如下：

```
<el-alert title="成功提示的文案" type="success" show-icon center effect="dark"></el-alert>
```

（6）带有辅助性文字介绍。Element 除了为警告提供了标题，还为警告提供了更为详细的带辅助性文字介绍，在 el-alert 标签中设置 description 属性可以使警告带有辅助性文字介绍。代码如下：

```
<el-alert title="消息提示的文案" type="info" description="这是一条消息" effect="dark" style="height:
40px;"></el-alert>
```

（7）带有 icon 和辅助性文字介绍。结合带有 icon 的样式和带有辅助性文字介绍的样式可以实现更为复杂的警告样式，只需要在 el-alert 标签中设置 show-icon 属性和 description 属性就可以使警告既带有 icon，也带有辅助性文字介绍。代码如下：

```
<el-alert title="警告提示的文案" type="warning" description="这是一条警告" effect="dark"style="height:
40px;"show-icon></el-alert>
```

5.2 加载（Loading）

Element 为加载提供了不同的样式与用法，本节将从区域加载、自定义、整页加载等方面讲解加载的样式与用法。

需要注意的是，加载并不是一个组件，而是与其他组件相结合使用的一个属性。本节将结合表格组件来对加载样式进行讲解。由于表格组件所占物理空间较大，因此本节分别用一个页面布局对每个样式进行讲解。

（1）区域加载。区域加载是指局部空间的加载，Element 为加载提供了两种方法，分别是指令和服务。对于指令方法，Element 是通过 v-loading 属性来实现加载的，若 v-loading 的取值为 true，则处于加载状态，否则处于正常状态，本节的示例均使用指令方法进行讲解。

本示例首先创建 Vue 实例，data 中的 tableData 是一个对象数组，每个对象都表示表格中的一行数据，vectorSource 是一个用于显示文本标注的全局变量，v-loading 属性用于控制是否显示加载状态，methods 中的 handleRowClick 事件是表格的行单击事件；然后结合地图显示区域加载的样式（见图 5-3）。

在本示例中，单击表格中的某一行，使表格处于加载状态，经过 2 s 后在地图上显示相应行的文本标注，表格的加载状态消失并恢复正常状态，如图 5-4 所示。代码如下：

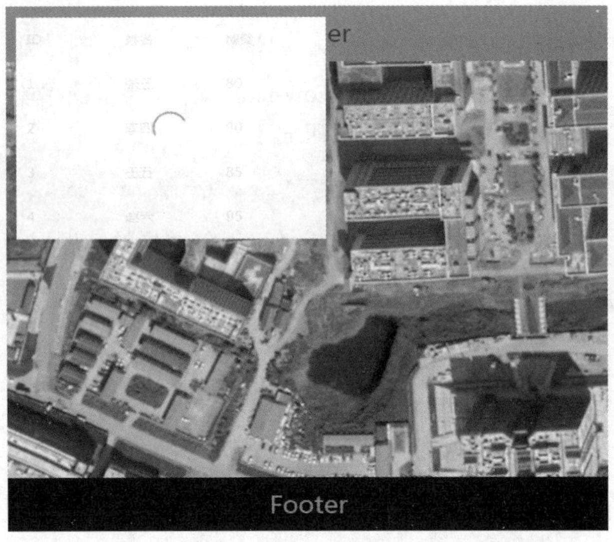

图 5-3　区域加载的样式

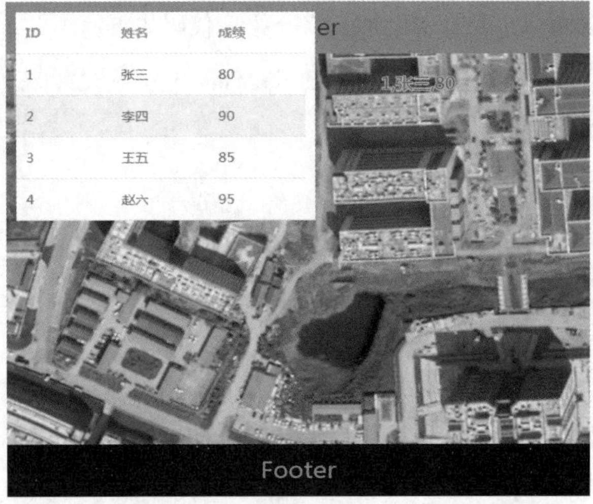

图 5-4　结合 OpenLayers 实现地图的文本标注

```
var vm = new Vue({
    el: "#app",
    data:
    {
        loading: false,
        tableData: [{
            ID: 1,
            name: "张三",
            grade: 80

        }, {
            ID: 2,
```

```
            name: "李四",
            grade: 90
        }, {
            ID: 3,
            name: "王五",
            grade: 85
        }, {
            ID: 4,
            name: "赵六",
            grade: 95
        }],
        //矢量标注的数据源
        vectorSource: new ol.source.Vector()
    },
    methods: {
        handleRowClick: function (val) {
            //创建矢量标注样式函数
            var createLabelStyle = function (feature) {
                return new ol.style.Style({
                    text: new ol.style.Text({
                        //位置
                        textAlign: 'center',
                        //基准线
                        textBaseline: 'middle',
                        //文字样式
                        font: 'normal 18px  微软雅黑',
                        //文本内容
                        text: feature.get('name'),
                        //文本填充样式（即文字颜色）
                        fill: new ol.style.Fill({ color: '#aa4400' }),
                        stroke: new ol.style.Stroke({ color: '#ffcc33', width: 2 })
                    })
                });
            }
            //创建点的坐标
            var pt = [2750, 1900];
            //实例化 Vector 要素，通过矢量图层添加到地图容器中
            var iconFeature = new ol.Feature({
                geometry: new ol.geom.Point(pt),
                name: val.ID + "," + val.name + "," + val.grade
            })
            iconFeature.setStyle(createLabelStyle(iconFeature));
            //添加文本标注之前先清空原来的标注
            this.vectorSource.clear();
            this.vectorSource.addFeature(iconFeature);
            //矢量标注图层
            var vectorLayer = new ol.layer.Vector({
```

```
                source: this.vectorSource
            });
            //表格显示区域加载状态
            this.loading = true;
            //定义一个定时器，2 s 后在地图上显示文本标注，表格的加载状态消失并恢复正常状态
            setTimeout(() => {
                map.addLayer(vectorLayer);
                this.loading=false;
                console.log(this.loading)
            }, 2000)
        }
    }
})
```

区域加载样式对应的标签代码如下：

```
<el-table :data="tableData" @row-click="handleRowClick" v-loading="loading" >
    <el-table-column prop="ID" label="ID" width="100">
    </el-table-column>
    <el-table-column prop="name" label="姓名" width="100">
    </el-table-column>
    <el-table-column prop="grade" label="成绩" width="100">
    </el-table-column>
</el-table>
```

（2）自定义。通过设置 element-loading-text、element-loading-spinner 和 element-loading-spinner 属性可以自定义加载时显示的文本内容、加载的图标和背景色。

本示例首先创建 Vue 实例，本示例创建的 Vue 实例与区域加载创建的 Vue 实例一样，这里不再给出代码；然后结合地图展示自定义的样式，如图 5-5 所示。

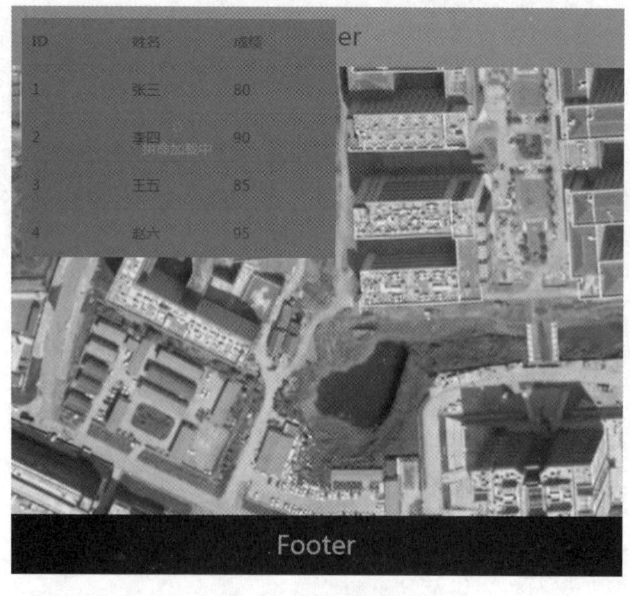

图 5-5　自定义的样式

自定义样式对应的标签代码如下：

```
<el-table :data="tableData" @row-click="handleRowClick" v-loading="loading" element-loading-text="拼命
加载中" element-loading-spinner="el-icon-loading"     element-loading-background="rgba(0, 0, 0, 0.6)">
    <el-table-column prop="ID" label="ID" width="100">
    </el-table-column>
    <el-table-column prop="name" label="姓名" width="100">
    </el-table-column>
    <el-table-column prop="grade" label="成绩" width="100">
    </el-table-column>
</el-table>
```

（3）整页加载。在 v-loading 属性中添加 fullscreen 修饰符可以实现整页加载，此时的加载页面会遮盖整个界面。若想要锁定屏幕的滚动，可以使用 lock 修饰符。

本示例首先创建 Vue 实例，本示例创建的 Vue 实例与区域加载创建的 Vue 实例一样，这里不再给出代码；然后结合地图展示整页加载的样式，如图 5-6 所示。

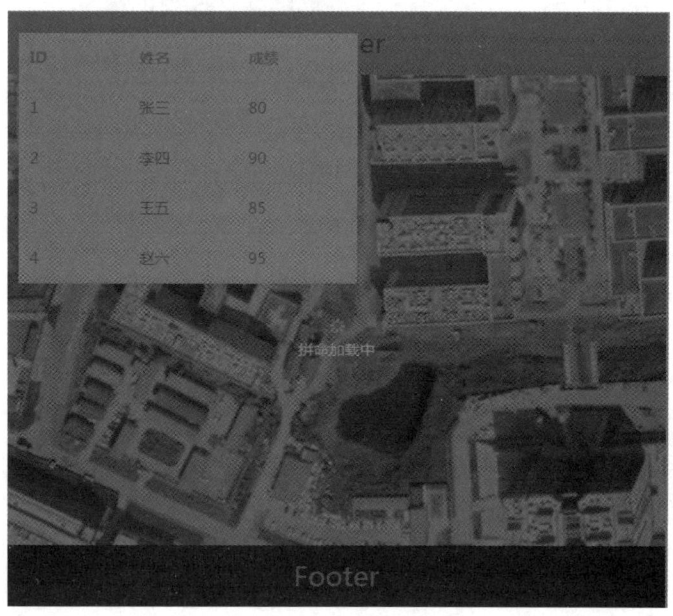

图 5-6　整页加载的样式

整页加载样式对应的标签代码如下：

```
<el-table :data="tableData" @row-click="handleRowClick"
v-loading.fullscreen.lock="loading" element-loading-text="拼命加载中"
element-loading-spinner="el-icon-loading"
element-loading-background="rgba(0, 0, 0, 0.6)">
    <el-table-column prop="ID" label="ID" width="100">
    </el-table-column>
    <el-table-column prop="name" label="姓名" width="100">
    </el-table-column>
    <el-table-column prop="grade" label="成绩" width="100">
```

```
    </el-table-column>
</el-table>
```

5.3 消息提示（Message）

Element 为消息提示提供了不同的样式与用法，本节将从基础用法、不同状态、可关闭、文字居中等方面讲解消息提示的样式与用法。本节首先结合地图展示与消息提示相关的按钮布局，如图 5-7 所示；然后对消息提示的样式与用法进行讲解。

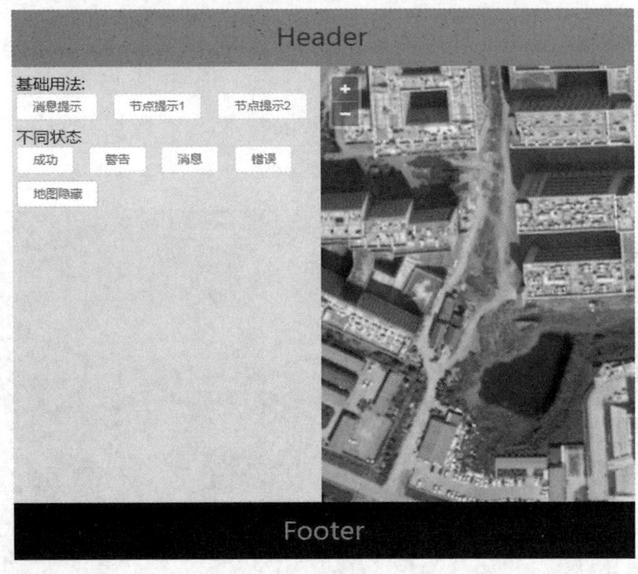

图 5-7　与消息提示相关的按钮布局

由于消息提示是通过事件触发的，因此本示例首先给出按钮布局的代码，然后对消息提示的样式与用法进行讲解。代码如下：

```
<!-- 1.基础用法 -->
<el-row>
    <div>基础用法:</div>
    <el-button :plain="true" @click="openInfo" size="mini">消息提示</el-button>
    <el-button :plain="true" @click="openNode" size="mini">节点提示 1</el-button>
    <el-button :plain="true" @click="openHTML" size="mini">节点提示 2</el-button>
</el-row>
<!-- 2.不同状态、可关闭、文字居中 -->
<el-row>
    <div>不同状态</div>
    <el-button :plain="true" @click="open1" size="mini">成功</el-button>
    <el-button :plain="true" @click="open2" size="mini">警告</el-button>
    <el-button :plain="true" @click="open3" size="mini">消息</el-button>
    <el-button :plain="true" @click="open4" size="mini">错误</el-button>
</el-row>
```

```
<el-row>
    <el-button :plain="true" @click="handleHideMap" size="mini">地图隐藏</el-button>
</el-row>
```

（1）基础用法。消息提示常用于主动操作后的反馈提示。Element 注册了一个$message 方法，该方法既可以接收一个字符串，也可以接收一个 VNode（VNode 是 JavaScript 对象，VNode 表示 Virtual DOM，用 JavaScript 对象来描述真实的 DOM，可以把 DOM 标签的属性和内容变成对象的属性），还可以将一个对象当成参数，这些参数会被显示为正文的内容。分别单击基础用法中的“消息提示”按钮、“节点提示 1”按钮和“节点提示 2”按钮，消息提示便会展示出来，消息提示默认停留 3 s 便会消失，如图 5-8 所示。代码如下：

图 5-8　基础用法的样式

```
//单击“消息提示”按钮时的响应事件
openInfo: function () {
    this.$message("这是一条消息提示")
},
//单击“节点提示 1”按钮时的响应事件
openNode: function () {
    var ele = this.$createElement;
    this.$message({
        message: ele("el-button", null, "我是按钮")
    })
},
//单击“节点提示 2”按钮时的响应事件
openHTML: function () {
    this.$message({
        dangerouslyUseHTMLString: true,
        message: '<h1>我是一级标题</h1>'
    })
},
```

（2）不同状态。在$message 方法将一个对象当成参数时，可以在该对象中设置 type 属性来控制不同的状态，showClose 属性用来控制消息提示是否可关闭，center 属性用来控制消息提示的内容是否居中显示。不同状态包括“成功”“警告”“消息”“错误”按钮，单击不同的按钮，消息提示便会展示出来，消息提示默认停留 3 s 便会消失，如图 5-9 所示。代码如下：

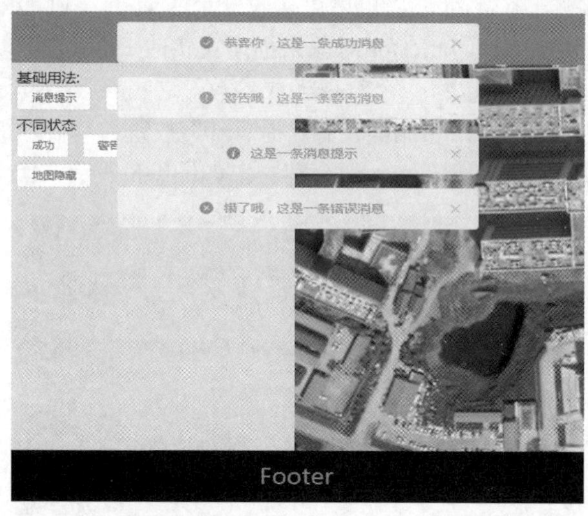

图 5-9　不同状态的样式

```
//单击“成功”按钮时的响应事件
open1: function () {
    this.$message({
        showClose: true,
        center: true,
        message: '恭喜你，这是一条成功消息',
        type: 'success'
    });
},
//单击“警告”按钮时的响应事件
open2: function () {
    this.$message({
        showClose: true,
        center: true,
        message: '警告哦，这是一条警告消息',
        type: 'warning'
    });
},
//单击“消息”按钮时的响应事件
open3: function () {
    this.$message({
        showClose: true,
        center: true,
        message: '这是一条消息提示',
        type: "info"
```

```
        })
    },
    //单击“错误”按钮时的响应事件
    open4: function () {
        this.$message({
            showClose: true,
            center: true,
            message: '错了哦，这是一条错误消息',
          type: 'error'
      });
```

（3）地图隐藏。地图隐藏的样式如图 5-10 所示，代码较为简单，本示例不再给出相应的代码。

图 5-10　地图隐藏的样式

5.4　弹框（MessageBox）

Element 为弹框提供了不同的样式与用法，下面我们将从消息提示、确认消息、提交内容、自定义、使用 HTML、取消与关闭等几个方面进行讲解。本节首先结合地图展示与弹框相关的按钮布局，如图 5-11 所示；然后对弹框的样式与用法进行讲解。

由于消息弹框主要是通过事件来触发的，因此本示例首先给出了按钮布局的代码，然后对弹框的样式与用法进行讲解。代码如下：

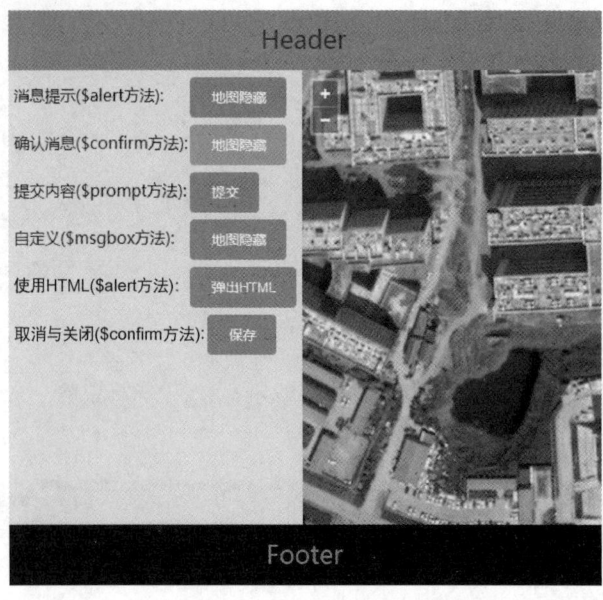

图 5-11　与弹框相关的按钮布局

```
<!-- 1.消息提示 -->
<el-row>
    <el-col :span="15">消息提示($alert 方法):</el-col>
    <el-col :span="9">
        <el-button type="primary" @click="open1">地图隐藏</el-button>
    </el-col>
</el-row>
<!-- 2.确认消息 -->
<el-row>
    <el-col :span=15>确认消息($confirm 方法):</el-col>
    <el-col :span=9>
        <el-button type="primary" @click="open2">地图隐藏</el-button>
    </el-col>
</el-row>
<!-- 3.提交内容 -->
<el-row>
    <el-col :span="15">提交内容($prompt 方法):</el-col>
    <el-col :span=9>
        <el-button type="primary" @click="open3">提交</el-button>
    </el-col>
</el-row>
<!-- 4.自定义 -->
<el-row>
    <el-col :span="15">自定义($msgbox 方法):</el-col>
    <el-col :span="9">
        <el-button type="primary" @click="open4">地图隐藏</el-button>
    </el-col>
```

```
</el-row>
<!-- 5.使用 HTML -->
<el-row>
    <el-col :span="15">使用 HTML:</el-col>
    <el-col :span="9">
        <el-button type="primary" @click="open5">弹出 HTML</el-button>
    </el-col>
</el-row>
<!-- 6.取消与关闭 -->
<el-row>
    <el-col :span="15">取消与关闭:</el-col>
    <el-col :span="9">
        <el-button type="primary" @click="open6">保存</el-button>
    </el-col>
</el-row>
```

（1）消息提示。Element 是通过对$alert、$confirm、$prompt、$msgbox 方法进行注册来实现弹框的，其中$alert、$confirm、$prompt 是基于$msgbox 方法进行再次包装。在用户进行操作时会触发消息提示，消息提示会中断用户的操作，直到用户确认提示的消息后才可关闭。本示例通过$alert 方法来模拟系统的弹框，用户无法通过按下 ESC 按键或单击弹框外来关闭弹框。$alert 方法的第一个参数是弹框的内容，第二个参数是弹框的标题，第三个参数是一个对象。在本示例中，对象（第三个参数）通过 confirmButtonText 属性来控制“确定”按钮的文本内容，通过 center 属性来控制弹框内容是否居中，该对象还通过一个回调函数来展示 5.3 节介绍的消息提示。单击消息提示对应的“地图隐藏”按钮，可弹出弹框，单击弹框中的“确定”按钮可隐藏地图，并弹出消息提示，如图 5-12 和图 5-13 所示。代码如下：

图 5-12　消息提示的样式（一）

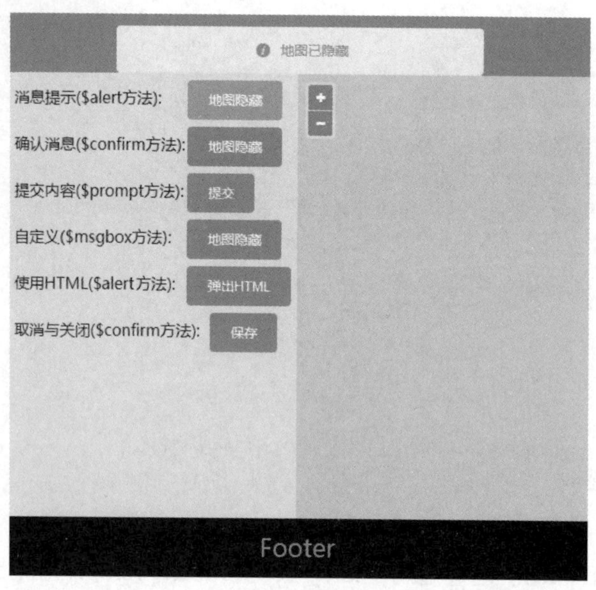

图 5-13　消息提示的样式（二）

```
open1: function () {
    this.$alert('确定隐藏地图？', '提示', {
        center: true,
        confirmButtonText: '确定',
        callback: (action) => {
            //imageLayer 图层隐藏
            imageLayer.setVisible(false);
            this.$message({
                center: true,
                type: 'info',
                message: '地图已隐藏'
            });
        }
    })
},
```

（2）确认消息。确认消息的作用是提示用户确认其进行的操作，并询问进行该操作时是否会用到对应的弹框。本示例通过$confirm 方法来模拟系统的确认。$confirm 方法的第一个参数是弹框的内容，第二个参数是弹框的标题，第三个参数是一个对象。在本示例中，对象（第三个参数）通过 confirmButtonText 属性来控制“确定”按钮的文本内容，通过 ancelButtonText 属性来控制“取消”按钮的文本内容，通过 type 属性来控制弹框内容的显示类型。本示例结合 5.3 节介绍的消息提示并通过 Promise 来处理后续的响应。当单击确认消息所对应的“地图隐藏”按钮时，可弹出弹框，单击弹框中的“确定”按钮可隐藏地图，并弹出消息提示，如图 5-14 和图 5-15 所示。代码如下：

图 5-14　确认消息的样式（一）

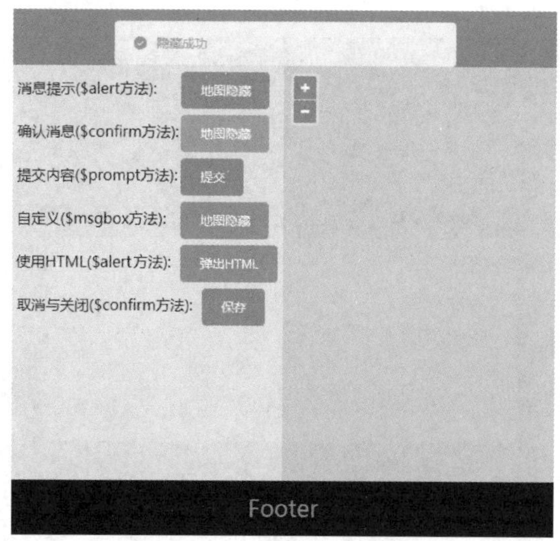

图 5-15　确认消息的样式（二）

```
open2: function () {
    this.$confirm("确定隐藏地图？ ", "提示", {
        confirmButtonText: '确定',
        cancelButtonText: '取消',
        type: 'warning',
    }).then(() => {
        imageLayer.setVisible(false);
        this.$message({
            type: "success",
            message: "隐藏成功"
```

```
            })
        }).catch(() => {
            this.$message({
                type: "info",
                message: "已取消隐藏"
            })
        })
    }
```

（3）提交内容。在用户进行操作时会触发提交内容，提交内容会中断用户的操作，提示用户在输入框中输入相关的内容。本示例通过$prompt 方法来模拟系统的提交。$prompt 方法的第一个参数是弹框的内容，第二个参数是弹框的标题，第三个参数是一个对象。在本示例中，对象（第三个参数）通过 confirmButtonText 属性来控制“确定”按钮的文本内容，通过 ancelButtonText 属性来控制“取消”按钮的文本内容，通过 inputPattern 属性来规定输入框的匹配模式，通过 inputErrorMessage 属性来处理匹配模式错误的情况。本示例结合 5.3 节介绍的消息提示并通过 Promise 来处理后续的响应。当单击提交内容对应的“提交”按钮时，可弹出弹框，在输入框中输入手机号码后单击“确定”按钮可弹出消息提示，如图 5-16 和图 5-17 所示。代码如下：

图 5-16　提交内容的样式（一）

```
open3: function () {
    this.$prompt("请输入手机号", "提示", {
        confirmButtonText: '确定',
        cancelButtonText: '取消',
        inputPattern: /^1(3|4|5|7|8)\d{9}$/,
        inputErrorMessage: '手机号格式不正确'
    }).then((obj) => {
        this.$message({
            type: 'success',
```

```
                message: '你的手机号是: ' + obj.value
            });
        }).catch(() => {
            this.$message({
                type: 'info',
                message: '取消输入'
            });
        })
    },
```

图 5-17 提交内容的样式（二）

（4）自定义。自定义用于配置不同的内容。本示例调用了$msgbox 方法，该方法的参数是一个对象。在本示例中，对象（$msgbox 方法的参数）通过 title 属性来控制弹框的标题，通过 message 属性来控制弹框的内容，通过 showCancelButton 属性来控制是否显示“取消”按钮，通过 confirmButtonText 来控制“确认”按钮的文本内容，通过 ancelButtonText 属性来控制“取消”按钮的文本内容。本示例还使用了 beforeClose 属性，该属性的值是一个方法，该方法会在弹框关闭之前被调用，同时暂停了弹框的关闭功能。beforeClose 属性的值接收三个参数，分别为 action、instance 和 done，其中 instance 为弹框实例本身；done 是一个方法，若需要关闭弹框，则需要调用 done 方法。beforeClose 属性的值能够在弹框关闭前对弹框实例进行一些操作，如单击“确定”按钮添加加载状态等。在单击自定义对应的“地图隐藏”按钮时，可弹出弹框，单击弹框中的“确定”按钮后，使“确定”按钮处于加载状态（变为“执行中...”按钮），持续 3 s 后隐藏地图，如图 5-18 和图 5-19 所示。代码如下：

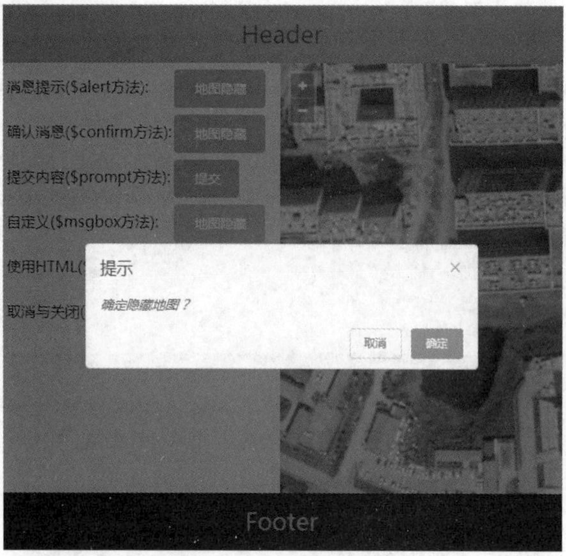

图 5-18　自定义的样式（一）

图 5-19　自定义的样式（二）

```
open4: function () {
    const ele = this.$createElement;
    this.$msgbox({
        title: "提示",
        message: ele('i', null, "确定隐藏地图？"),
        showCancelButton: true,
        confirmButtonText: '确定',
        cancelButtonText: '取消',
        beforeClose: (action, instance, done) => {
            if (action === "confirm") {
                //显示加载组件
```

```
                instance.confirmButtonLoading = true;
                instance.confirmButtonText = '执行中...';
                setTimeout(() => {
                    //imageLayer 图层隐藏
                    imageLayer.setVisible(false);
                    instance.confirmButtonLoading = false;
                    done()
                }, 3000);
            }
            else {
                done();
            }
        }
    }).then(() => {
        this.$message({
            type: 'success',
            message: "隐藏成功"
        })
    }).catch(() => {
        this.$message({
            type: 'info',
            message: "已取消隐藏"
        })
    })
},
```

（5）使用 HTML。$alert、$confirm 和$prompt 方法都支持传入 HTML 片段，只需要将 dangerouslyUseHTMLString 属性设置为 true，就可以直接将 HTML 片段添加到弹框中。本示例调用的是$alert 方法，当单击使用 HTML 对应的“弹出 HTML”按钮时，可弹出弹框，该弹框的内容是 HTML 片段中的一级标题，如图 5-20 所示。代码如下：

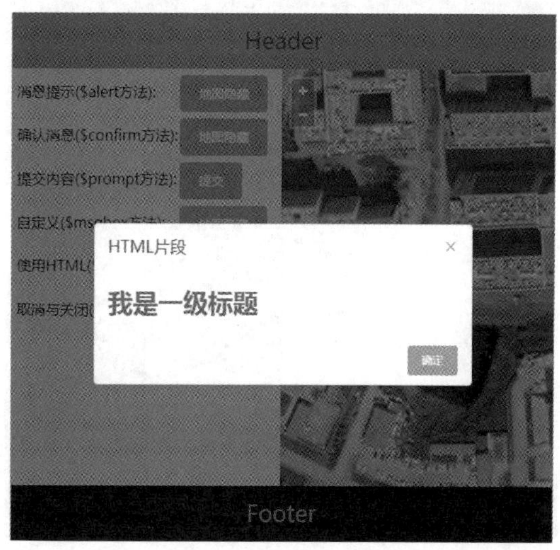

图 5-20　使用 HTML 的样式

```
open5: function () {
    this.$alert('<h1>我是一级标题<h1>', "HTML 片段", {
        dangerouslyUseHTMLString: true
    })
},
```

（6）取消与关闭。在默认情况下，当用户触发取消事件（如单击“取消”按钮）或触发关闭事件（如单击“关闭”按钮、单击遮罩层或按下 ESC 按键）时，Promise 的 reject 会调用回调函数 callbac，触发这两种事件的回调函数的参数都是 cancel。但在某些场景下，取消事件和关闭事件有着不同的含义。如果将 distinguishCancelAndClose 属性设置为 true，则上述两种事件的回调函数参数分别为 cancel 和 close。在本实例中，当单击取消与关闭所对应的“保存”按钮时，可弹出弹框，如图 5-21 所示；当单击弹框中的“放弃”按钮和“保存”按钮时，会弹出两个不同的消息提示框，分别如图 5-22 和图 5-23 所示。代码如下：

图 5-21　取消与关闭的样式（一）

图 5-22　取消与关闭的样式（二）

图 5-23　取消与关闭的样式（三）

```
open6: function () {
    this.$confirm('是否保存？', '确认信息', {
        distinguishCancelAndClose: true,
        confirmButtonText: '保存',
        cancelButtonText: '放弃'
    }).then(() => {
        this.$message({
            type: 'info',
            message: '保存修改'
        });
    }).catch((action) => {
        this.$message({
            type: action === "cancel" ? "warning":"info",
            message: action === "cancel" ? '放弃保存并离开页面' : '停留在当前页面'
        })

    })
}
```

5.5 通知（Notification）

Element 为通知提供了不同的样式与用法，本节将从基础用法、带有倾向性和自定义位置、带有偏移、隐藏关闭按钮等方面讲解通知的样式与用法。本节首先结合地图展示与通知相关的按钮布局，如图 5-24 所示；然后对通知的样式与用法进行讲解。

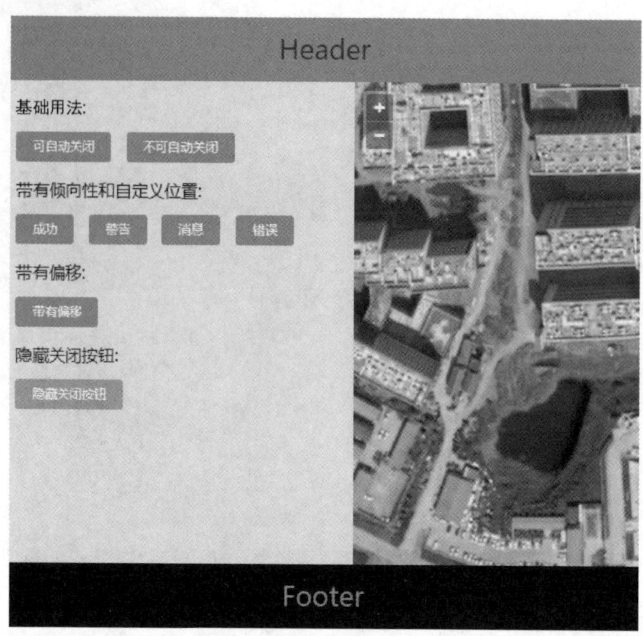

图 5-24　与通知相关的按钮布局

由于通知是通过事件来触发的，因此本示例首先给出了按钮布局的代码，然后对通知的样式与用法进行讲解。代码如下：

```
<!-- 1.基础用法 -->
<el-row>
    <div>基础用法:</div>
    <el-button type="primary" @click="open1_1" size="mini">可自动关闭</el-button>
    <el-button type="primary" @click="open1_2" size="mini">不可自动关闭</el-button>
</el-row>
<!-- 2.带有倾向性和自定义位置 -->
<el-row>
    <div>带有倾向性和自定义位置:</div>
    <el-button type="primary" @click="open2_1" size="mini">成功</el-button>
    <el-button type="primary" @click="open2_2" size="mini">警告</el-button>
    <el-button type="primary" @click="open2_3" size="mini">消息</el-button>
    <el-button type="primary" @click="open2_4" size="mini">错误</el-button>
</el-row>
<!-- 3.带有偏移 -->
<el-row>
    <div>带有偏移:</div>
    <el-button type="primary" @click="open3" size="mini">带有偏移</el-button>
</el-row>
<!-- 4.隐藏关闭按钮 -->
<el-row>
    <div>隐藏关闭按钮:</div>
    <el-button type="primary" @click="open4" size="mini">隐藏关闭按钮</el-button>
</el-row>
```

（1）基础用法。消息提示多用于主动操作后的反馈提示，而通知多用于被动提醒。Element 注册了$notify 方法，该方法接收的参数是一个对象，在本示例中，对象（$notify 方法的参数）通过 title 属性来控制通知的标题；通过 message 属性来控制通知的内容；通过 duration 属性来设置通知在页面的停留时间，单位为 ms，若停留时间为 0，则该通知不会自动关闭。在默认情况下，通知会出现在页面的右上角，停留 4.5 s 后消失。在本示例中，单击基础用法对应的各个按钮，会显示相应的通知，如图 5-25 所示。代码如下：

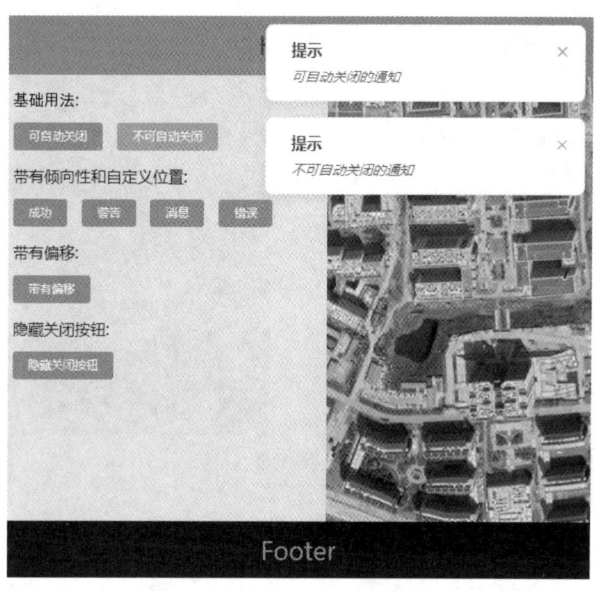

图 5-25 基础用法的样式

```
//单击“可自动关闭”按钮时的响应事件
open1_1: function () {
    var ele = this.$createElement;
    this.$notify({
        title: "提示",
        message: ele("i", { style: "color:red" }, "可自动关闭的通知")
    })
},
//单击“不可自动关闭”按钮时的响应事件
open1_2: function () {
    var ele = this.$createElement;
    this.$notify({
        title: "提示",
        message: ele("i", { style: "color:red" }, "不可自动关闭的通知"),
        duration: 0
    })
},
```

（2）带有倾向性和自定义位置。Element 注册了$notify 方法，该方法接收的参数是一个对象。在本示例中，对象（$notify 方法的参数）通过 title 属性来控制通知的标题；通过 message

属性来控制通知的内容；通过 type 属性来控制通知的类型，该属性的取值可以是 success、warning、info、error；通过 position 属性来控制通知的位置，该属性的取值可以是 top-right、top-left、bottom-right、bottom-left，默认的取值是 top-right。在本示例中，分别单击带有倾向性和自定义位置对应的各个按钮，可以显示相应的通知，如图 5-26 和图 5-27 所示。代码如下：

图 5-26 带有倾向性和自定义位置的样式（一）

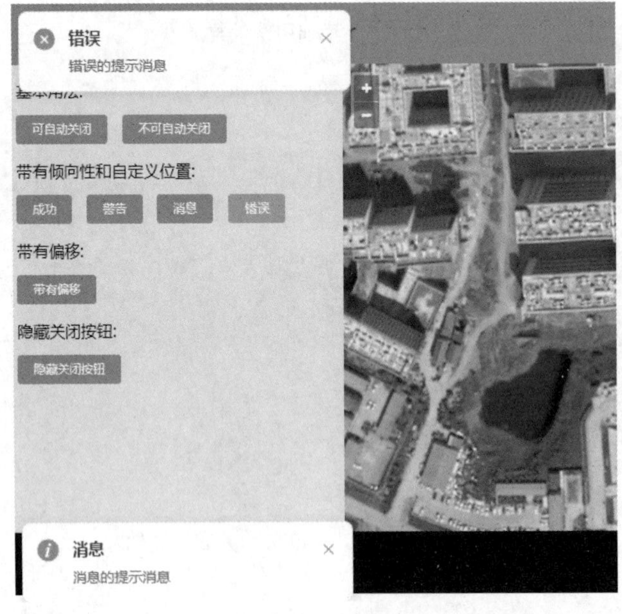

图 5-27 带有倾向性和自定义位置的样式（二）

```
//单击“成功”按钮时的响应事件
open2_1: function () {
    this.$notify({
```

```
            title: "成功",
            message: "成功的提示消息",
            type: "success"
        })
    },
    //单击“警告”按钮时的响应事件
    open2_2: function () {
        this.$notify({
            title: "警告",
            message: "警告的提示消息",
            type: "warning",
            position: 'bottom-right'
        })
    },
    //单击“消息”按钮时的响应事件
    open2_3: function () {
        this.$notify({
            title: "消息",
            message: "消息的提示消息",
            type: "info",
            position: 'bottom-left'
        })
    },
    //单击“错误”按钮时的响应事件
    open2_4: function () {
        this.$notify({
            title: "错误",
            message: "错误的提示消息",
            type: "error",
            position: 'top-left'
        })
    },
```

（3）带有偏移。Element 注册了$notify 方法，该方法接收的参数是一个对象。在本示例中，对象（$notify 方法的参数）通过 title 属性来控制通知的标题；通过 message 属性来控制通知的内容；通过 type 属性来控制通知的类型；通过 offset 属性来控制通知的偏移量，单位为像素。在本示例中，单击带有偏移对应的按钮，可以显示相应的通知，如图 5-28 所示。代码如下：

```
    //单击“带有偏移”按钮时的响应事件
    open3: function () {
        this.$notify({
            title: '偏移',
            message: '这是一条带有偏移的提示消息',
            offset: 200
        })
    },
```

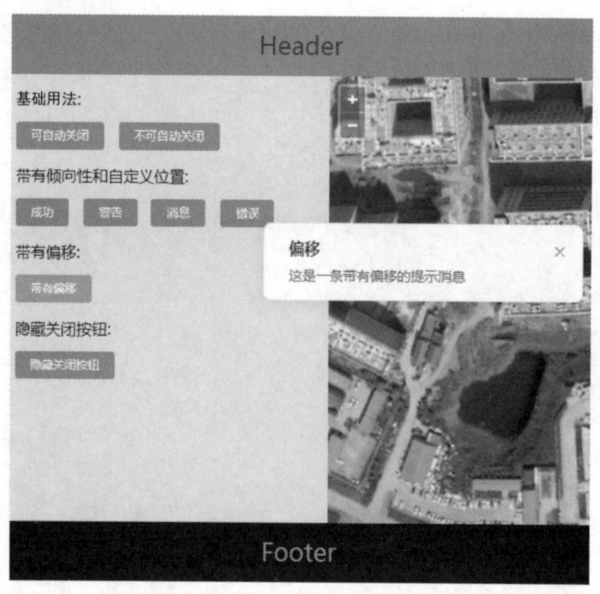

图 5-28　带有偏移的样式

（4）隐藏关闭按钮。Element 注册了$notify 方法，该方法接收的参数是一个对象。在本示例中，对象（$notify 方法的参数）通过 title 属性来控制通知的标题；通过 message 属性来控制通知的内容；通过 dangerouslyUseHTMLString 属性来控制通知是否接收 HTML；通过 showClose 属性来控制通知是否显示“关闭”按钮。在本示例中，单击隐藏关闭对应的按钮，可以显示相应的通知，如图 5-29 所示。代码如下：

图 5-29　隐藏关闭按钮的样式

```
//单击“隐藏关闭按钮”时的响应事件
open4: function () {
    this.$notify({
```

```
            title:"HTML 的一级标题",
            message: '<h1>这是一级标题</h1>',
            dangerouslyUseHTMLString: true,
            showClose:false
        })
    },
```

5.6 思考与练习题

深刻理解各个通知组件的区别与联系，分别实现各个组件的弹出效果。

第 6 章 导航组件

Element 为导航组件提供了非常丰富、详细的样式，本章将结合 Vue 和 OpenLayers 对 Element 的各个导航组件进行详细的介绍。

6.1 导航菜单（NavMenu）

Element 为导航菜单提供了不同的样式与用法，本节将从顶栏、侧栏、折叠等方面讲解导航菜单的样式与用法。由于导航菜单所占的物理空间较大，因此本节分别用一个页面布局对每个样式进行讲解。

（1）顶栏。Element 通过对 el-menu、el-submenu、el-menu-item 等的封装提供了导航菜单，其中 el-menu 为父标签，el-menu-item 为子标签。若需要在菜单中生成二级菜单，则需要使用 el-submenu 标签。需要注意的是，可通过 template 标签中的 slot 属性来设置二级标题。导航菜单的默认模式是垂直模式，将 el-menu 标签中的 mode 属性设置为 horizontal，可以使导航菜单的模式变成水平模式。通过 el-menu 标签中的 default-active 属性可以控制默认激活的子菜单，通过 el-menu 标签中的 background-color 属性可以控制导航菜单的背景色，通过 el-menu 标签中的 text-color 属性可以控制导航菜单的文字颜色，通过 el-menu 标签中的 active-text-color 属性可以控制导航菜单某个元素被单击后的显示颜色。

本示例首先创建 Vue 实例，data 中的 activeIndex 为当前默认激活的子菜单的索引号，methods 中的 handleSelect 是导航菜单被选择时的响应事件；然后结合地图展示顶栏的样式，如图 6-1 所示。在本示例中，单击一号食堂、二号食堂、教工食堂，地图的视角将会发生变化。代码如下：

```
var vm = new Vue({
    el: "#app",
    data:
    {
        activeIndex: "1-2"
    },
    methods: {
```

```
            //导航菜单被选择时的响应事件
            handleSelect(key, keyPath) {
                switch(key) {
                    case "1-1-1":
                    map.getView().setCenter([2700, 1600]);
                    map.getView().setZoom(4);
                    break;
                    case "1-1-2":
                    map.getView().setCenter([3000, 2000]);
                    map.getView().setZoom(4);
                    break;
                    case "1-1-3":
                    map.getView().setCenter([3500, 1200]);
                    map.getView().setZoom(4);
                    break;
                }

            }
        }
    })
```

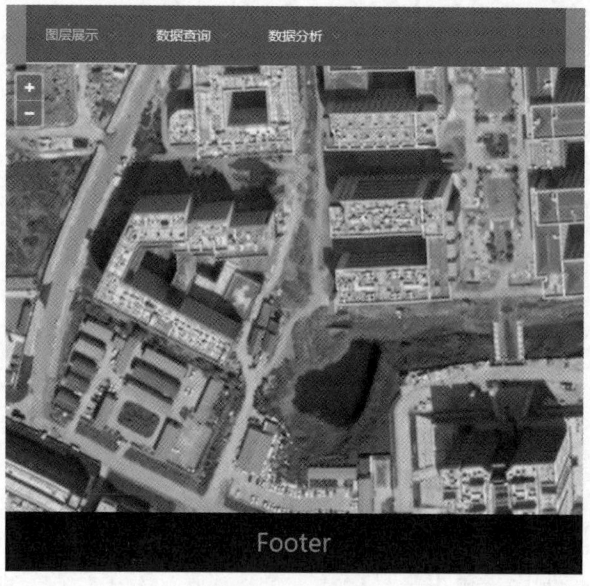

图 6-1　顶栏的样式

顶栏样式对应的标签代码如下：

```
    <el-menu  mode="horizontal"  :default-active="activeIndex"  @select="handleSelect"  background-color=
"#545c64" text-color="#fff" active-text-color="#ffd04b">
        <el-submenu index="1">
            <template slot="title">图层展示</template>
            <el-submenu index="1-1">
                <template slot="title">食堂</template>
```

```
            <el-menu-item index="1-1-1">一号食堂</el-menu-item>
            <el-menu-item index="1-1-2">二号食堂</el-menu-item>
            <el-menu-item index="1-1-3">教工食堂</el-menu-item>
        </el-submenu>
        <el-menu-item index="1-2">教学楼</el-menu-item>
        <el-menu-item index="1-3">图书馆</el-menu-item>
    </el-submenu>
    <el-submenu index="2">
        <template slot="title">数据查询</template>
        <el-menu-item index="2-1">空间查询</el-menu-item>
        <el-menu-item index="2-2">属性查询</el-menu-item>
    </el-submenu>
    <el-submenu index="3">
        <template slot="title">数据分析</template>
        <el-menu-item index="3-1">图表分析</el-menu-item>
        <el-menu-item index="3-2">路径分析</el-menu-item>
        <el-menu-item index="3-3">热力图分析</el-menu-item>
    </el-submenu>
</el-menu>
```

（2）侧栏。通过在 template 标签中嵌入 i 标签可以在导航菜单中添加图表。本示例首先创建 Vue 实例，data 和 methods 为空；然后结合地图展示侧栏的样式，如图 6-2 所示。代码如下：

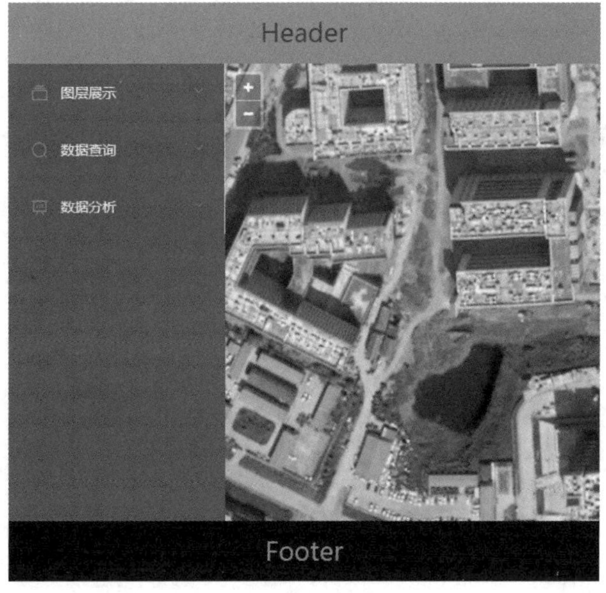

图 6-2　侧栏的样式

```
var vm = new Vue({
    el: "#app",
    data: {
    },
```

```
        methods: {
        }
    })
```

侧栏样式对应的标签代码如下：

```
<el-menu    background-color="#545c64"
    text-color="#fff" active-text-color="#ffd04b">
    <el-submenu index="1">
        <template slot="title">
            <i class="el-icon-files"></i>
            <span>图层展示</span>
        </template>
        <el-submenu index="1-1">
            <template slot="title">食堂</template>
            <el-menu-item index="1-1-1">一号食堂</el-menu-item>
            <el-menu-item index="1-1-2">二号食堂</el-menu-item>
            <el-menu-item index="1-1-3">教工食堂</el-menu-item>
        </el-submenu>
        <el-menu-item index="1-2">教学楼</el-menu-item>
        <el-menu-item index="1-3">图书馆</el-menu-item>
    </el-submenu>
    <el-submenu index="2">
        <template slot="title">
            <i class="el-icon-search"></i>
            <span>数据查询</span>
        </template>
        <el-menu-item index="2-1">空间查询</el-menu-item>
        <el-menu-item index="2-2">属性查询</el-menu-item>
    </el-submenu>
    <el-submenu index="3">
        <template slot="title">
            <i class="el-icon-data-analysis"></i>
            <span>数据分析</span>
        </template>
        <el-menu-item index="3-1">图表分析</el-menu-item>
        <el-menu-item index="3-2">路径分析</el-menu-item>
        <el-menu-item index="3-3">热力图分析</el-menu-item>
    </el-submenu>
</el-menu>
```

（3）折叠。通过 el-menu 标签中的 collapse 属性可以控制导航菜单是否为折叠状态。本示例首先创建 Vue 实例，data 中的 isCollapse 属性与 el-menu 标签中的 collapse 属性、el-radio-group 标签中的 v-model 属性相关联，即通过按钮样式的单选框来控制导航菜单是否为折叠状态；然后结合地图展示折叠的样式，如图 6-3 所示。代码如下：

图 6-3　折叠的样式

```
var vm = new Vue({
    el: "#app",
    data: {
        isCollapse:true
    },
    methods: {

    }
})
```

折叠样式对应的标签代码如下：

```
<el-radio-group v-model="isCollapse">
    <el-radio-button :label="false">展开</el-radio-button>
    <el-radio-button :label="true">收起</el-radio-button>
</el-radio-group >
<el-menu :collapse="isCollapse">
    <el-submenu index="1">
        <template slot="title">
            <i class="el-icon-files"></i>
            <span>图层展示</span>
        </template>
        <el-submenu index="1-1">
            <template slot="title">食堂</template>
            <el-menu-item index="1-1-1">一号食堂</el-menu-item>
            <el-menu-item index="1-1-2">二号食堂</el-menu-item>
            <el-menu-item index="1-1-3">教工食堂</el-menu-item>
        </el-submenu>
```

```
            <el-menu-item index="1-2">教学楼</el-menu-item>
            <el-menu-item index="1-3">图书馆</el-menu-item>
        </el-submenu>
        <el-submenu index="2">
            <template slot="title">
                <i class="el-icon-search"></i>
                <span>数据查询</span>
            </template>
            <el-menu-item index="2-1">空间查询</el-menu-item>
            <el-menu-item index="2-2">属性查询</el-menu-item>
        </el-submenu>
        <el-submenu index="3">
            <template slot="title">
                <i class="el-icon-data-analysis"></i>
                <span>数据分析</span>
            </template>
            <el-menu-item index="3-1">图表分析</el-menu-item>
            <el-menu-item index="3-2">路径分析</el-menu-item>
            <el-menu-item index="3-3">热力图分析</el-menu-item>
        </el-submenu>
    </el-menu>
```

6.2 标签页（Tabs）

Element 为标签页提供了不同的样式与用法，本节将从基础用法、选项卡样式、卡片化、位置、自定义标签页、动态增减标签页等方面讲解标签页的样式与用法。由于物理空间的限制，因此本节用两个页面布局对标签页的样式进行讲解，如图 6-4 和图 6-5 所示。

图 6-4　标签页的样式（一）

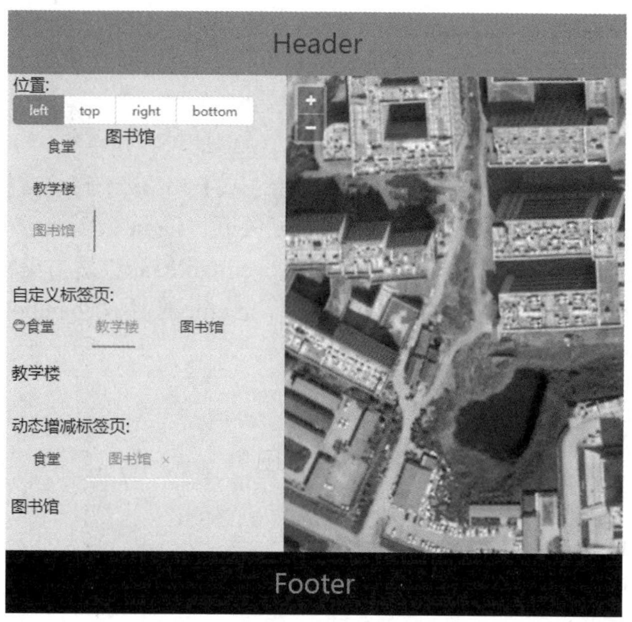

图 6-5　标签页的样式（二）

对于图 6-4，本节创建了 Vue 实例，其中 data 中的 name1、name2、name3 分别与三个标签页样式中 el-tabs 标签中的 v-model 相关联；methods 中的 handleClick 是基础用法中标签被单击时的响应事件，单击不同的标签时地图视图发生不同的变化。代码如下：

```
var vm = new Vue({
    el: "#app",
    data: {
        name1: "restaurant",
        name2: "teachBuilding",
        name3: "library"
    },
    methods: {
        //标签被单击时的响应事件
        handleClick: function (tab) {
            switch (tab.label) {
                case "食堂":
                    map.getView().setCenter([2700, 1600]);
                    map.getView().setZoom(4);
                    break;
                case "图书馆":
                    map.getView().setCenter([3000, 2000]);
                    map.getView().setZoom(4);
                    break;
                case "教学楼":
                    map.getView().setCenter([3500, 1200]);
                    map.getView().setZoom(4);
                    break;
```

```
                }
            }
        }
    })
```

对于图 6-5，本节创建了 Vue 实例，其中 data 中的 position 与位置样式中 el-radio-group 标签中的 v-model、el-tabs 标签中的 tab-position 相关联，name4、name5、name6 分别与三个标签页样式中 el-tabs 标签中的 v-model 相关联，editableTabs 是渲染标签页的对象数组，newTableIndex 是一个用于创建新标签索引的变量，methods 中的 handleEdit 是动态增减标签页样式中标签被添加或移除时的响应事件。

```
    var vm = new Vue({
        el: "#app",
        data: {
            position: "left",
            name4: "library",
            name5: "teachBuilding",
            name6: "library",
            editableTabs: [{
                title: "食堂",
                name: "restaurant",
            }, {
                title: "图书馆",
                name: "library",
            }],
            newTableIndex: 0
        },
        methods: {
            /*单击标签页中的标签被添加或移除时的响应事件，该事件的处理方法接收两个参数，第一个参数是当前标签的 name，第二个参数是当前执行的响应事件。若 action 为 add，则 targetName 为 null；若 action 为 remove，则 targetName 为当前单击标签的名字*/
            handleEdit: function (targetName, action) {
                //添加标签
                if (action === 'add') {
                    //通过递增的索引号来为每一个新增的标签定义唯一的 name 和 title
                    this.newTableIndex++;
                    var newTabName = "newTab" + this.newTableIndex;
                    var newTabTitle="新标签"+this.newTableIndex
                    //在 editableTabs 数组中添加一个新的标签对象
                    this.editableTabs.push({
                        title: newTabTitle,
                        name: newTabName,
                    })
                    //当前激活的标签名字为新增标签的名字
                    this.name6 = newTabName;
                }
                //移除标签
```

```
            if (action === 'remove') {
                //遍历 editableTabs 数组中所有的标签
                this.editableTabs.forEach((value,index)=>{
                    //找到将要移除的标签
                    if(value.name===targetName){
                        /*获取将要移除标签的邻近标签，若待移除的标签是最后一个标签，则取其左边的标签为邻近标签，否则取其右边的标签为邻近标签。当然代码也可以写为
                        this.editableTabs[index+1]||this.editableTabs[index-1];*/
                        var nextTab=(index+1===this.editableTabs.length)?
                                        this.editableTabs[index-1]:this.editableTabs[index+1]
                        //若邻近标签不为空，则当前激活的标签为邻近标签
                        if(nextTab){
                            this.name6=nextTab.name;
                        }
                    }
                })
                /*通过 filter 方法筛选 editableTabs 数组，将待移除的标签移除，得到一个新的数组*/
                this.editableTabs = this.editableTabs.filter(function (tab) {
                    return tab.name !== targetName
                });
            }


        }
    }
})
```

（1）基础用法。Element 通过向 el-tabs 标签中嵌入 el-tab-pane 标签来实现标签页。其中 el-tabs 标签中的 v-moedel 用于监听 el-tab-pane 标签中的 name 属性，el-tab-pane 标签中的 label 属性用于显示标签的文本内容，el-tab-pane 标签的文本内容用于显示被激活的标签。代码如下：

```
<el-tabs v-model="name1" @tab-click="handleClick">
    <el-tab-pane label="食堂" name="restaurant">食堂</el-tab-pane>
    <el-tab-pane label="教学楼" name="teachBuilding">教学楼</el-tab-pane>
    <el-tab-pane label="图书馆" name="library">图书馆</el-tab-pane>
</el-tabs>
```

（2）选项卡样式。在 el-tabs 标签中将 type 属性设置为 card 即可实现选项卡样式。代码如下：

```
<el-tabs v-model="name2" type="card">
    <el-tab-pane label="食堂" name="restaurant">食堂</el-tab-pane>
    <el-tab-pane label="教学楼" name="teachBuilding">教学楼</el-tab-pane>
    <el-tab-pane label="图书馆" name="library">图书馆</el-tab-pane>
</el-tabs>
```

（3）卡片化。在 el-tabs 标签中将 type 属性设置为 border-card 即可实现卡片化样式。代码如下：

```
<el-tabs v-model="name2" type="border-card">
    <el-tab-pane label="食堂" name="restaurant">食堂</el-tab-pane>
    <el-tab-pane label="教学楼" name="teachBuilding">教学楼</el-tab-pane>
    <el-tab-pane label="图书馆" name="library">图书馆</el-tab-pane>
</el-tabs>
```

（4）位置。通过 el-tabs 标签中的 tab-position 属性可以控制不同标签页的位置，该属性的取值可以为 left、right、top、bottom，默认的取值为 top。代码如下：

```
<el-radio-group v-model="position" size="mini">
    <el-radio-button label="left">left</el-radio-button>
    <el-radio-button label="top">top</el-radio-button>
    <el-radio-button label="right">right</el-radio-button>
    <el-radio-button label="bottom">bottom</el-radio-button>
</el-radio-group>
<el-tabs v-model="name4" :tab-position="position">
    <el-tab-pane label="食堂" name="restaurant">食堂</el-tab-pane>
    <el-tab-pane label="教学楼" name="teachBuilding">教学楼</el-tab-pane>
    <el-tab-pane label="图书馆" name="library">图书馆</el-tab-pane>
</el-tabs>
```

（5）自定义标签页。自定义标签可以通过在 el-tab-pane 标签中嵌入其他标签来实现更加复杂的样式，例如可以在该标签中嵌入图标样式。代码如下：

```
<el-tabs v-model="name5">
    <el-tab-pane><span slot="label"><i class="el-icon-food">食堂</i></span>食堂</el-tab-pane>
    <el-tab-pane label="教学楼" name="teachBuilding">教学楼</el-tab-pane>
    <el-tab-pane label="图书馆" name="library">图书馆</el-tab-pane>
</el-tabs>
```

（6）动态增减标签页。动态增减标签页中的按钮只能在选项卡中使用，首先在 el-tabs 标签中添加 editable 属性，然后通过 edit 事件来实现标签页的动态增减。代码如下：

```
<el-tabs v-model="name6" type="card" editable @edit="handleEdit">
    <el-tab-pane v-for="(item,index) of editableTabs" :label="item.title" :name="item.name">
        {{item.title}}</el-tab-pane>
</el-tabs>
```

6.3 面包屑（Breadcrumb）

Element 为面包屑提供了不同的样式与用法，本节将从基础用法、图标分隔符两个方面讲解面包屑的样式与用法。本示例首先创建 Vue 实例，由于本示例比较简单，data 和 methods 都为空；然后结合地图展示面包屑的样式，如图 6-6 所示。代码如下：

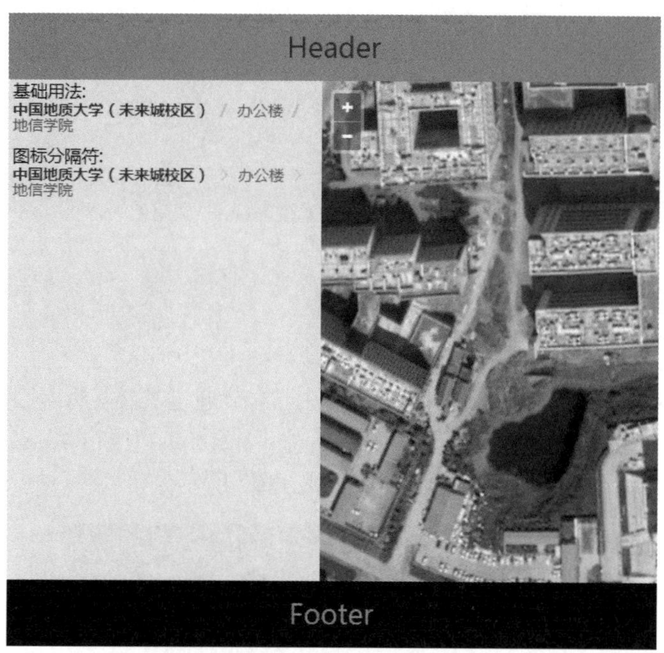

图 6-6　面包屑的样式

```
var vm = new Vue({
    el: "#app",
    data: {
    },
    methods:{

    }
})
```

（1）基础用法。面包屑是指显示当前页面的路径，通过面包屑可以快速返回之前的任意页面。Element 提供了 el-breadcrumb 标签和 el-breadcrumb-item 标签，通过在 el-breadcrumb 标签中使用 el-breadcrumb-item 标签，可以表示从首页开始的每一级路径；通过 el-breadcrumb 标签中的 separator 属性可以决定分隔符，分隔符只能是字符串，默认为斜杠；el-breadcrumb-item 标签中的 to 属性为路由跳转对象。由于本示例是个单页面，因此这里不展示页面跳转功能。代码如下：

```
<el-breadcrumb separator="/">
    <el-breadcrumb-item :to="{ path: '/' }">首页</el-breadcrumb-item>
    <el-breadcrumb-item><a href="/">活动管理</a></el-breadcrumb-item>
    <el-breadcrumb-item>活动列表</el-breadcrumb-item>
    <el-breadcrumb-item>活动详情</el-breadcrumb-item>
</el-breadcrumb>
```

（2）图标分隔符。通过 el-breadcrumb 标签中的 separator-class 属性可以使用图标作为分隔符。代码如下：

```
<el-breadcrumb separator-class="el-icon-arrow-right">
    <el-breadcrumb-item :to="{ path: '/' }">首页</el-breadcrumb-item>
    <el-breadcrumb-item>活动管理</el-breadcrumb-item>
    <el-breadcrumb-item>活动列表</el-breadcrumb-item>
    <el-breadcrumb-item>活动详情</el-breadcrumb-item>
</el-breadcrumb>
```

6.4 页头（PageHeader）

Element 为页头提供了一种样式，即基础用法，本节通过具体的示例讲解基础用法的样式与用法。本示例首先创建 Vue 实例，其中 data 为空，methods 中的 goBack 是单击“返回”时的响应事件；然后结合地图展示页头的样式，如图 6-7 所示。

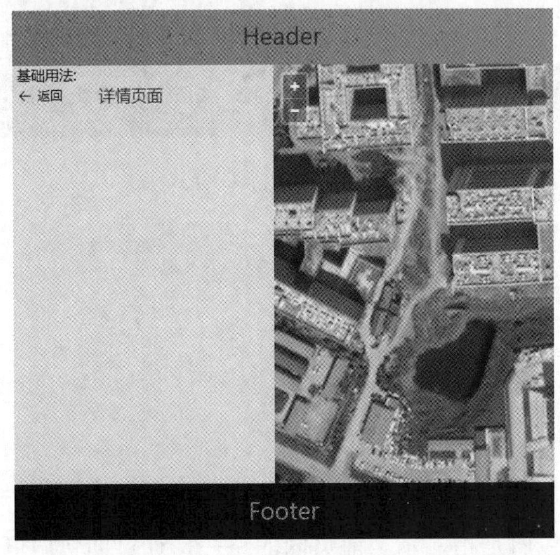

图 6-7　页头的样式

由于本示例只有一个页面，因此本示例结合消息提示的用法弹出一个提示框，如图 6-8 所示。代码如下：

```
var vm = new Vue({
    el: "#app",
    data: {
    },
    methods:{
        //单击“返回”时的响应事件
        goBack:function(){
            this.$message("返回")
        }
    }
})
```

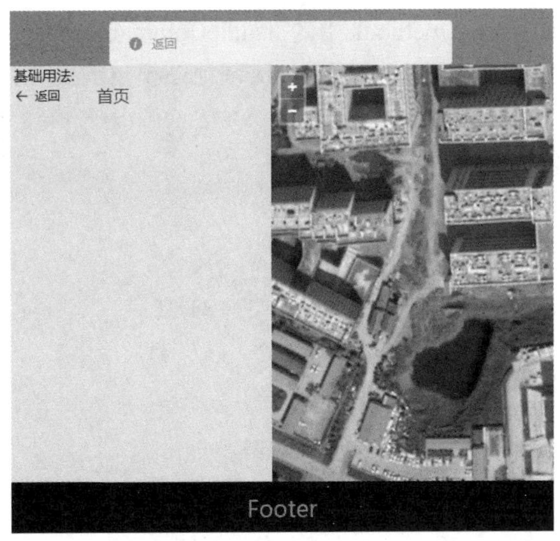

图 6-8 结合消息提示的用法弹出一个提示框

如果页面的路径比较简单，则推荐使用页头而非面包屑。Element 通过对 el-page-header 标签的封装提供了页头组件，content 属性是要显示的文本内容。代码如下：

```
<el-page-header @back="goBack" content="详情页面"></el-page-header>
```

6.5 下拉菜单（Dropdown）

Element 为下拉菜单提供了不同的样式与用法，本节将从基础用法、触发对象、菜单隐藏方式、指令事件等方面讲解下拉菜单的样式与用法。本节首先结合地图展示下拉菜单的样式，如图 6-9 所示；然后对下拉菜单的样式与用法进行讲解。

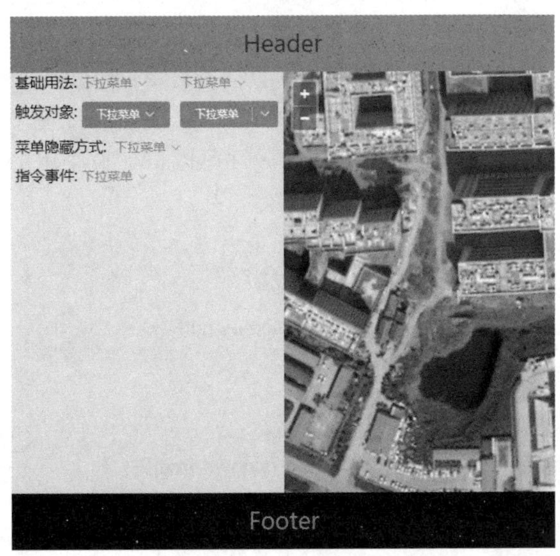

图 6-9 下拉菜单的样式

本示例首先创建 Vue 实例，methods 中的 handleCommand 是指令事件样式中对应下拉菜单被选择时的响应事件；然后结合 OpenLayers 实现地图视角的变化。代码如下：

```
var vm = new Vue({
    el: "#app",
    data: {
    },
    methods: {
        //指令事件样式中对应下拉菜单被选择时的响应事件
        handleCommand: function (command) {
            switch (command) {
                case "食堂":
                    map.getView().setCenter([2700, 1600]);
                    map.getView().setZoom(4);
                break;
                case "图书馆":
                    map.getView().setCenter([3000, 2000]);
                    map.getView().setZoom(4);
                break;
                case "教学楼":
                    map.getView().setCenter([3500, 1200]);
                    map.getView().setZoom(4);
                break;
            }
        }
    }
})
```

（1）基础用法。Element 是通过 el-dropdown、el-dropdown-menu 和 el-dropdown-item 标签来实现下拉菜单的。其中 el-dropdown 为父标签，el-dropdown-menu 为 el-dropdown 的子标签，el-dropdown-item 为 el-dropdown-menu 的子标签。在默认情况下，下拉菜单是通过 hover 方式展开的。若要通过 click 方式展开下拉菜单，则只需要在 el-dropdown 标签中将 trigger 属性设置为 click 即可。在 el-dropdown-item 标签中添加 divided 属性可以使下拉菜单选项之间出现分割线，在 el-dropdown-item 标签中添加 disabled 属性可以将下拉菜单的选项设置为禁止选择状态。代码如下：

```
<el-dropdown>
    <span class="el-dropdown-link">
        下拉菜单<i class="el-icon-arrow-down el-icon--right"></i>
    </span>
    <el-dropdown-menu slot="dropdown">
        <el-dropdown-item>食堂</el-dropdown-item>
        <el-dropdown-item divided>图书馆</el-dropdown-item>
        <el-dropdown-item disabled>教学楼</el-dropdown-item>
    </el-dropdown-menu>
</el-dropdown>
```

```
<el-dropdown trigger="click">
    <span class="el-dropdown-link">
        下拉菜单<i class="el-icon-arrow-down el-icon--right"></i>
    </span>
    <el-dropdown-menu slot="dropdown">
        <el-dropdown-item>食堂</el-dropdown-item>
        <el-dropdown-item>图书馆</el-dropdown-item>
        <el-dropdown-item>教学楼</el-dropdown-item>
    </el-dropdown-menu>
</el-dropdown>
```

（2）触发对象。通过按钮来触发下拉菜单的方式有两种：第一种方式是直接在 el-dropdown 标签中嵌入 el-button 标签；第二种方式是在 el-dropdown 标签中添加属性 split-button。在采用第二种方式时，按钮分为两个部分，左边表示按钮的功能，右边用于触发下拉菜单。另外，通过 el-dropdown 标签中的 size 属性可以控制按钮的尺寸，size 属性的取值为 large、medium、small、mini，默认的取值为 large。代码如下：

```
<el-dropdown>
    <el-button type="primary" size="mini">
        下拉菜单<i class="el-icon-arrow-down el-icon--right"></i>
    </el-button>
    <el-dropdown-menu slot="dropdown">
        <el-dropdown-item>食堂</el-dropdown-item>
        <el-dropdown-item>图书馆</el-dropdown-item>
        <el-dropdown-item>教学楼</el-dropdown-item>
    </el-dropdown-menu>
</el-dropdown>
<el-dropdown split-button type="primary" size="mini">
    下拉菜单
    <el-dropdown-menu slot="dropdown">
        <el-dropdown-item>食堂</el-dropdown-item>
        <el-dropdown-item>图书馆</el-dropdown-item>
        <el-dropdown-item>教学楼</el-dropdown-item>
    </el-dropdown-menu>
</el-dropdown>
```

（3）菜单隐藏方式。在默认情况下，单击下拉菜单选项后下拉菜单会隐藏起来，将 el-dropdown 标签中的 hide-on-click 属性设置为 false，可以不隐藏下拉菜单。代码如下：

```
<el-dropdown :hide-on-click="false">
    <span class="el-dropdown-link">
        下拉菜单<i class="el-icon-arrow-down el-icon--right"></i>
    </span>
    <el-dropdown-menu slot="dropdown">
        <el-dropdown-item>食堂</el-dropdown-item>
        <el-dropdown-item>图书馆</el-dropdown-item>
        <el-dropdown-item>教学楼</el-dropdown-item>
```

```
        </el-dropdown-menu>
    </el-dropdown>
```

（4）指令事件。在 el-dropdown 标签中绑定 command 方法，可以监听下拉菜单被选中的事件。代码如下：

```
    <el-dropdown @command="handleCommand">
        <span class="el-dropdown-link">
            下拉菜单<i class="el-icon-arrow-down el-icon--right"></i>
        </span>
        <el-dropdown-menu slot="dropdown">
            <el-dropdown-item command="食堂">食堂</el-dropdown-item>
            <el-dropdown-item command="图书馆">图书馆</el-dropdown-item>
            <el-dropdown-item command="教学楼">教学楼</el-dropdown-item>
        </el-dropdown-menu>
    </el-dropdown>
```

6.6 步骤条（Steps）

Element 为步骤条提供了不同的样式与用法，本节将从基础用法、有描述的步骤条、居中的步骤条、带图标的步骤条、竖式步骤条、简洁风格的步骤条等方面讲解步骤条的样式与用法。由于物理空间的限制，因此本节用两个页面布局对步骤条的样式与用法进行讲解，如图 6-10 和图 6-11 所示。

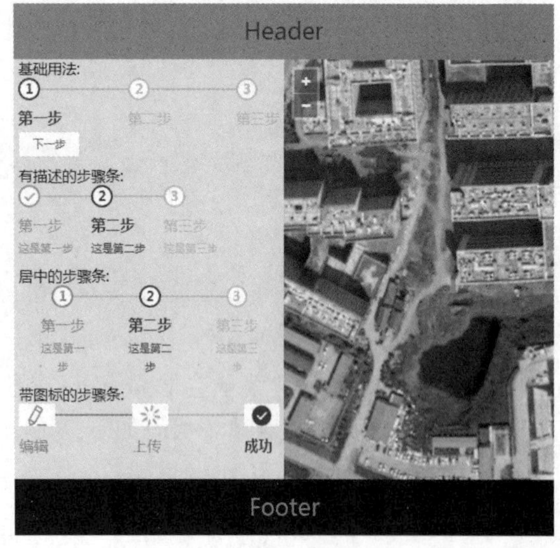

图 6-10　步骤条的样式（一）

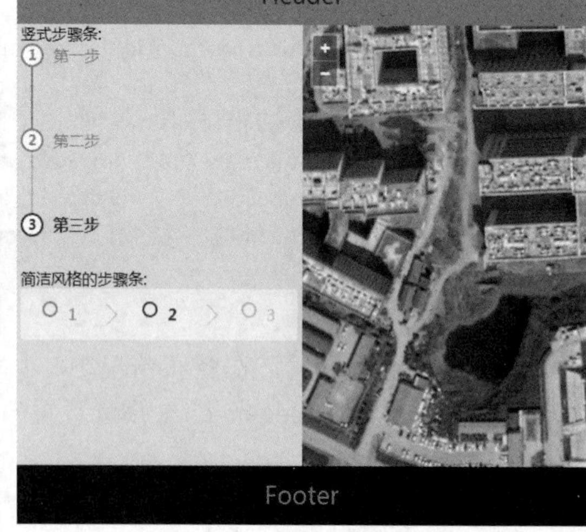

图 6-11　步骤条的样式（二）

对于图 6-10，本示例首先创建 Vue 实例，其中 data 中的 active 与基础用法中 el-steps 标签中的 v-model 属性相关联，methods 中的 next 方法是单击基础用法中“下一步”按钮时的响应事件；然后单击“下一步”按钮，每单击一次“下一步”按钮，地图的透明度都会发生变化，如图 6-11 所示。代码如下：

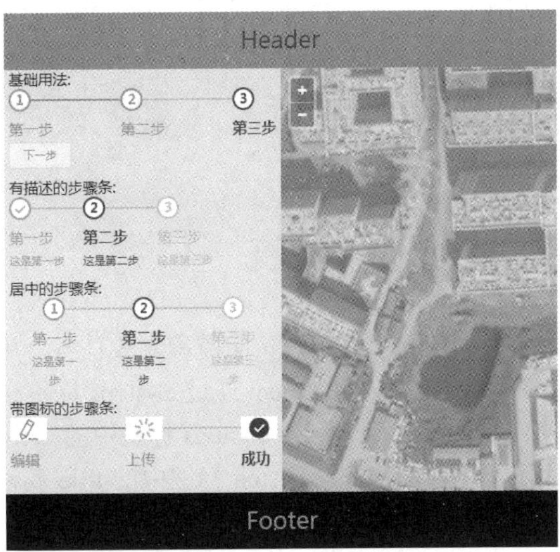

图 6-12　单击“下一步”按钮时地图透明度的变化

```
var vm = new Vue({
    el: "#app",
    data: {
        active: 0
    },
    methods: {
        //单击基础用法中“下一步”按钮时的响应事件
        next: function () {
            if (this.active < 3) {
                this.active++
            }
            else {
                this.active = 0;
            }
            switch (this.active) {
                case 1:
                    imageLayer.setOpacity(1);
                break;
                case 2:
                    imageLayer.setOpacity(0.5);
                break;
                case 3:
                    imageLayer.setOpacity(0);
                break;
            }
        }
    }
```

```
})
```

对于图 6-11，本示例创建了 Vue 实例，其中的 data 和 methods 为空。代码如下：

```
var vm = new Vue({
    el: "#app",
    data: {
    },
    methods: {
    }
})
```

（1）基础用法。步骤条的作用是引导用户按照流程完成任务，用户可根据实际应用场景设定步骤数量，步骤数量不得少于 2 步。Element 通过 el-steps 标签和 el-step 标签提供了步骤条，其中 el-steps 为父标签，el-step 为子标签。el-steps 标签中的 active 属性接收一个 Number 类型的值，从 0 开始，决定了当前的步骤。el-step 标签中的 title 属性用来控制步骤的标题。代码如下：

```
<el-steps :active="active" >
    <el-step title="第一步"></el-step>
    <el-step title="第二步"></el-step>
    <el-step title="第三步"></el-step>
</el-steps>
<el-button @click="next" size="mini">下一步</el-button>
```

（2）有描述的步骤条。有描述的步骤条是指步骤条除了有必要的标题，还对标题做了进一步的文字描述。在 el-step 标签中添加 description 属性可以实现有描述的步骤条，通过 el-steps 标签中的 finish-status 属性可以设置步骤条的当前状态，finish-status 属性的取值有 wait、progress、finish、error 和 success，默认的取值为 progress。另外，通过 el-steps 标签中的 space 属性可以设置每个步骤条之间的距离，单位为像素；若没有 space 属性，则各个步骤条将采用自适应的间距。代码如下：

```
<el-steps :active="1" :space="80"  finish-status="success">
    <el-step title="第一步" description="这是第一步"></el-step>
    <el-step title="第二步" description="这是第二步"></el-step>
    <el-step title="第三步" description="这是第三步"></el-step>
</el-steps>
```

（3）居中的步骤条。在 el-steps 标签中添加 align-center 属性可以使步骤条居中显示。代码如下：

```
<el-steps :active="1" align-center>
    <el-step title="第一步" description="这是第一步"></el-step>
    <el-step title="第二步" description="这是第二步"></el-step>
    <el-step title="第三步" description="这是第三步"></el-step>
</el-steps>
```

（4）带图标的步骤条。在 el-step 标签中设置 icon 属性可以为步骤条添加图标。代码如下：

```
<el-steps :active="2">
    <el-step title="编辑" icon="el-icon-edit"></el-step>
    <el-step title="上传" icon="el-icon-loading"></el-step>
    <el-step title="成功" icon="el-icon-success"></el-step>
</el-steps>
```

（5）竖式步骤条。将 el-steps 标签中的 direction 属性设置为 vertical，可以竖向显示步骤条。代码如下：

```
<el-steps :active="2" direction="vertical">
    <el-step title="第一步"></el-step>
    <el-step title="第二步"></el-step>
    <el-step title="第三步"></el-step>
</el-steps>
```

（6）简洁风格的步骤条。在 el-steps 标签中添加 simple 属性可以实现简洁风格的步骤条。需要注意的是，在简洁风格的步骤条中，align-center、description、direction、space 等属性都将失效。代码如下：

```
<el-steps :active="1"  simple>
    <el-step title="1"></el-step>
    <el-step title="2"></el-step>
    <el-step title="3"></el-step>
</el-steps>
```

6.7　思考与练习题

（1）参考 6.1 节，先设计三个名为图层管理、数据查询、数据分析的一级菜单，再为每个一级菜单设计一些二级菜单，最后添加地图来实现一个 GIS 系统的整体布局。

（2）参考 6.6 节，编写一个反映自己学习进度的步骤条。

第 7 章 其他组件

Element 除了提供了前几章介绍的组件，还提供了大量的其他组件，本章将结合 Vue 和 OpenLayers 来详细地介绍 Element 的其他组件。

7.1 对话框（Dialog）

Element 为对话框提供了不同的样式与用法，本节将从基础用法、自定义内容、嵌套对话框等方面讲解对话框的样式与用法。由于对话框所占的物理空间较大，因此本节分别用一个页面布局来讲解每个样式与用法。

（1）基础用法。Element 通过对 el-dialog 标签的封装提供了对话框，在 el-dialog 标签中，需要设置 visible.sync 属性，若该属性值为 true 则显示对话框，否则不显示对话框；该标签中的 title 属性用于自定义标题。el-dialog 标签内部可分为两个部分，即 body 和 footer，footer 需要具名为 footer 的 slot。在 el-dialog 标签中添加 center 属性可以使标题和底部区域居中，注意，对话框的内容不受 center 属性的影响。

本示例首先创建 Vue 实例，data 中的 dialogVisible 与 el-dialog 标签中的 visible.sync 属性相关联，methods 中的 handleclick 是单击“隐藏地图”按钮时的响应事件；然后结合地图展示基础用法的样式，如图 7-1 所示。在本示例中，单击“隐藏地图”按钮，可弹出对话框，单击对话框中的“确定”按钮，地图将会隐藏起来，如图 7-2 所示。代码如下：

```
var vm = new Vue({
    el: "#app",
    data: {
        dialogVisible:false
    },
    methods: {
        //单击“隐藏地图”按钮时的响应事件
        handleClick:function(){
            this.dialogVisible=false;
            imageLayer.setVisible(false);
        }
```

```
    }
})
```

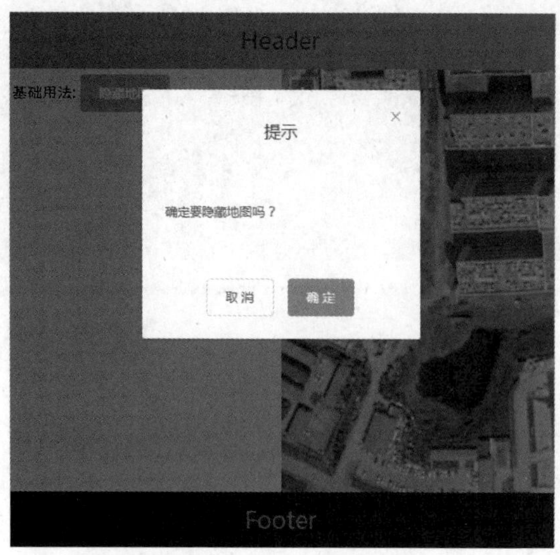

图 7-1　基础用法的样式

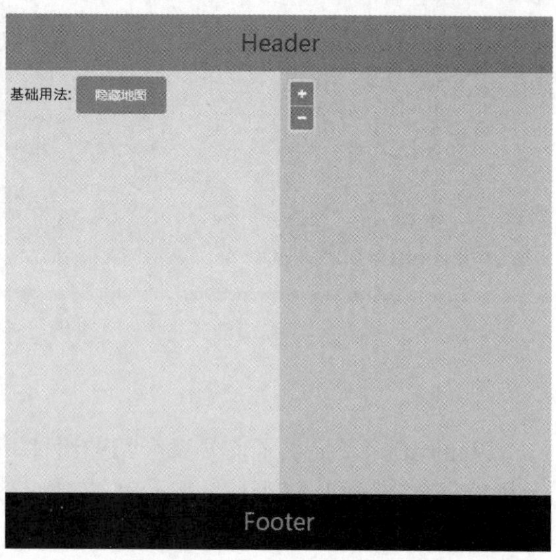

图 7-2　地图隐藏效果

基础用法样式对应的标签代码如下：

```
<el-button @click="dialogVisible = true" type="primary">隐藏地图</el-button>
<el-dialog  title="提示"  :visible.sync="dialogVisible" center>
    <span>确定要隐藏地图吗？</span>
    <span slot="footer">
        <el-button @click="dialogVisible = false">取 消</el-button>
        <el-button type="primary" @click="handleClick">确 定</el-button>
    </span>
</el-dialog>
```

（2）自定义内容。el-dialog 标签中的内容可以是任意的，甚至可以是表格。本示例首先创建 Vue 实例，data 中的 dialogVisible 与 el-dialog 标签中的 visible.sync 属性相关联，tableData 是表格渲染的数据，与 el-table 的 data 属性相关联；然后结合地图展示自定义内容的样式，如图 7-3 所示。在本示例中，单击“信息查询”按钮，可弹出具有表格的对话框。代码如下：

```
var vm = new Vue({
    el: "#app",
    data: {
        dialogVisible:false,
        tableData:[{
            ID:1,
            name:"张三",
            grade:80

        },{
            ID:2,
```

```
                name:"李四",
                grade:90
            },{
                ID:3,
                name:"王五",
                grade:85
            },{
                ID:4,
                name:"赵六",
                grade:95
            }],
        }
})
```

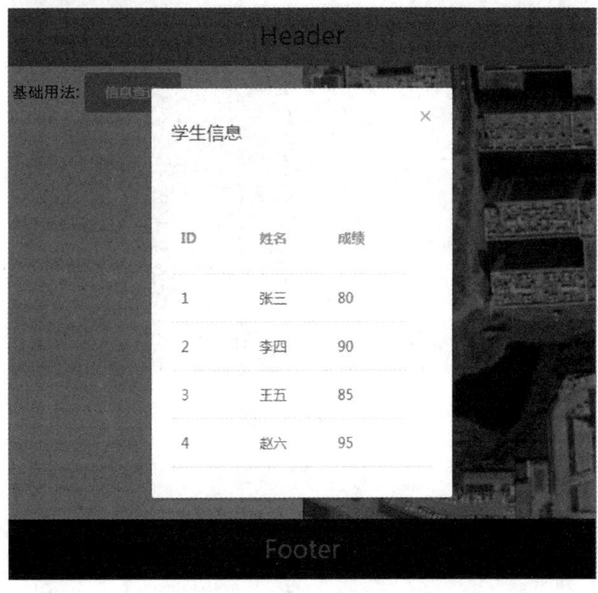

图 7-3　自定义内容的样式

自定义内容样式对应的标签代码如下：

```
<el-button @click="dialogVisible = true" type="primary">信息查询</el-button>
<el-dialog   title="学生信息"   :visible.sync="dialogVisible">
    <el-table :data="tableData">
        <el-table-column prop="ID" label="ID" width="80">
        </el-table-column>
        <el-table-column prop="name" label="姓名" width="80">
        </el-table-column>
        <el-table-column prop="grade" label="成绩" width="80">
        </el-table-column>
    </el-table>
</el-dialog>
```

（3）嵌套对话框。如果需要在一个对话框中嵌套另一个对话框，那么只需要先在父标签

el-dialog 内嵌入一个子标签 el-dialog，再在子标签 el-dialog 中添加 append-to-body 属性。

本示例首先创建 Vue 实例，data 中的 outerVisible 属性和 innerVisible 属性分别与父标签 el-dialog 和子标签 el-dialog 内的 visible.sync 属性相关联；然后结合地图展示嵌套对话框的样式，如图 7-4 所示。在本示例中，单击“打开对话框”按钮，可弹出一个对话框，在弹出的对话框中单击“打开内层对话框”按钮可打开内层对话框，第二个对话框便会显示出来。代码如下：

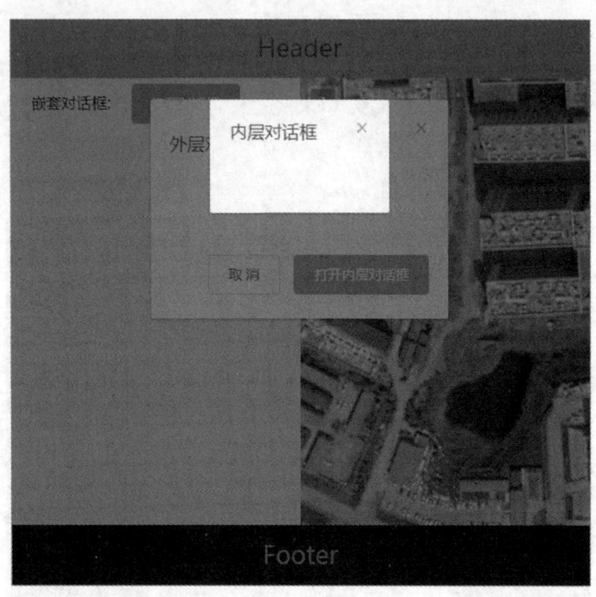

图 7-4　嵌套对话框的样式

```
var vm = new Vue({
    el: "#app",
    data: {
        outerVisible: false,
        innerVisible: false
    }
})
```

嵌套对话框样式对应的标签代码如下：

```
<el-button @click="outerVisible  = true" type="primary">打开对话框</el-button>
<el-dialog title="外层对话框" :visible.sync="outerVisible">
    <el-dialog title="内层对话框" width="30%" :visible.sync="innerVisible" append-to-body>
    </el-dialog>
    <div slot="footer">
        <el-button @click="outerVisible = false">取 消</el-button>
        <el-button type="primary" @click="innerVisible = true">打开内层对话框</el-button>
    </div>
</el-dialog>
```

7.2　文字提示（Tooltip）

Element 为文字提示提供了不同的样式与用法，本节将从基础用法、主题、更多内容、关闭文字提示等方面讲解文字提示的样式与用法。本节首先结合地图展示与文字提示相关的按钮布局，如图 7-5 所示；然后对文字提示的各个样式与用法进行讲解。

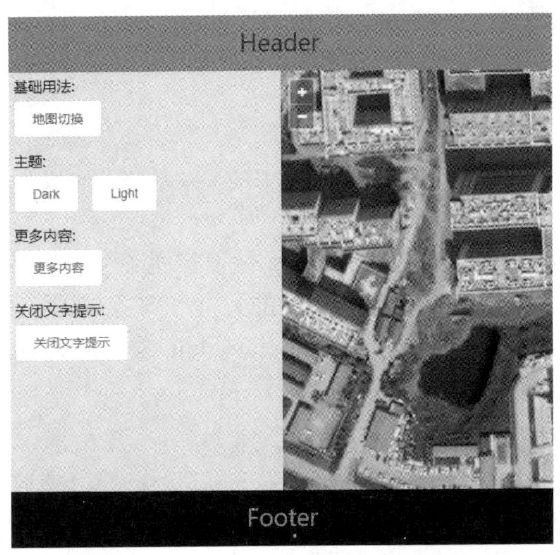

图 7-5　与文字提示相关的按钮布局

本示例首先创建 Vue 实例，methods 中的 handleClick 是单击基础用法中“地图切换”按钮时的响应事件；然后结合 OpenLayers 展示文字提示的样式，如图 7-6 所示。在本示例中，当鼠标光标移动到“地图切换”按钮上时，会显示文字提示；单击“地图切换”按钮，可实现地图隐藏或显示。代码如下：

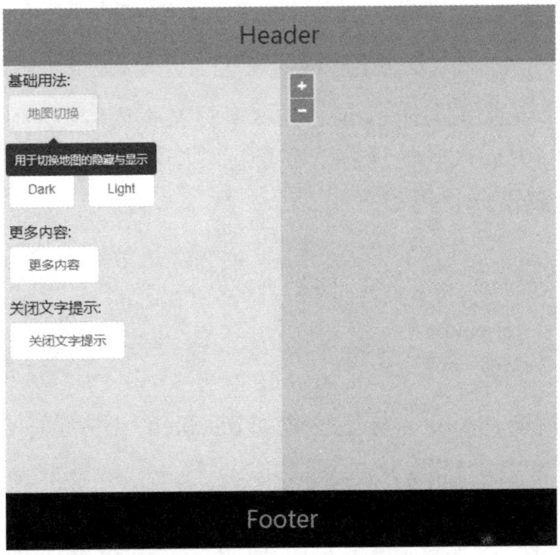

图 7-6　文字提示的样式

```
var vm = new Vue({
    el: "#app",
    data: {
    },
    methods: {
        //单击"地图切换"按钮时的响应事件
        handleClick:function(){
            //获取当前图层的可见性
            var visible = imageLayer.getVisible()
            //若地图当前可见，则变为不可见；若当前不可见，则变为可见
            imageLayer.setVisible(!visible);
        }
    }
})
```

（1）基础用法。Element通过封装el-tooltip标签实现了文字提示，其中content属性是文字提示的内容，placement属性决定了提示文字的方向和对齐位置，其取值包括top-start、top、top-end、bottom-start、bottom、bottom-end、left-start、left、left-end、right-start、right、right-end，默认的取值为bottom。代码如下：

```
<el-tooltip content="用于切换地图的隐藏与显示" placement="bottom">
    <el-button @click="handleClick">地图切换</el-button>
</el-tooltip>
```

（2）主题。Element为文字提示提供了两个不同的主题，通过el-tooltip标签中的effect属性可选择不同的主题，effect属性的取值包括dark和light，默认的取值是dark。代码如下：

```
<el-tooltip  effect="dark" content="Dark" placement="top">
    <el-button>Dark</el-button>
</el-tooltip>
<el-tooltip effect="light" content="Light" placement="right">
    <el-button>Light</el-button>
</el-tooltip>
```

（3）更多内容。如果要在文本提示框中展示多行文本内容或者设置文本内容的格式，则可以直接在el-tooltip标签中嵌入其他标签，在子标签中添加slot属性并赋值为content，用于替代tooltip中的content属性。代码如下：

```
<el-tooltip placement="bottom">
    <div slot="content">第一行<br />第二行</div>
    <el-button>更多内容</el-button>
</el-tooltip>
```

（4）关闭文字提示。在el-tooltip标签中添加disabled属性可关闭文字提示。代码如下：

```
<el-tooltip placement="bottom" content="内容" disabled>
    <el-button>关闭文字提示</el-button>
</el-tooltip>
```

7.3　弹出框（Popover）

Element 为弹出框提供了不同的样式与用法，本节将从基础用法、嵌套信息两个方面讲解弹出框的样式与用法。本节首先结合地图展示与弹出框相关的按钮布局，如图 7-7 所示；然后对弹出框的各个样式与用法进行讲解。

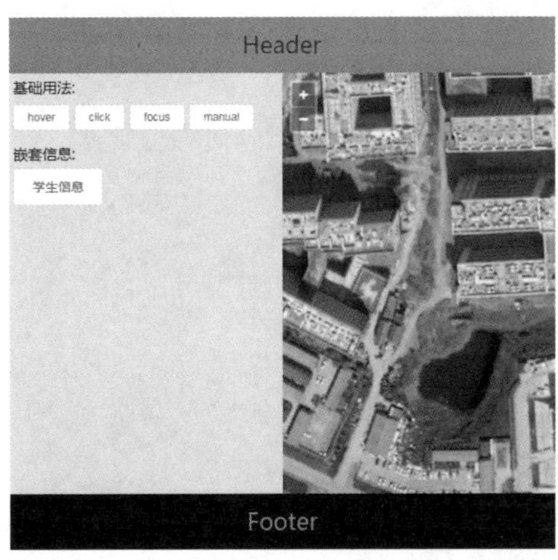

图 7-7　与弹出框相关的按钮布局

本示例首先创建 Vue 实例，data 中的 visible 与基础用法中第 4 个 el-popover 标签中的 v-model 属性相关联，tableData 与 el-table 标签中的 data 属性相关联，methods 中的 handleClick 是单击基础用法中“click”按钮时的响应事件，handleManualDblclick 是单击基础用法中“manual”按钮时的响应事件；然后结合 OpenLayers 展示弹出框的样式，如图 7-8 和图 7-9 所示。在本示例中，单击“click”按钮，可显示弹出框，同时地图的可见性会发生变化；单击“学生信息”按钮，可在弹出框中显示学生信息表。代码如下：

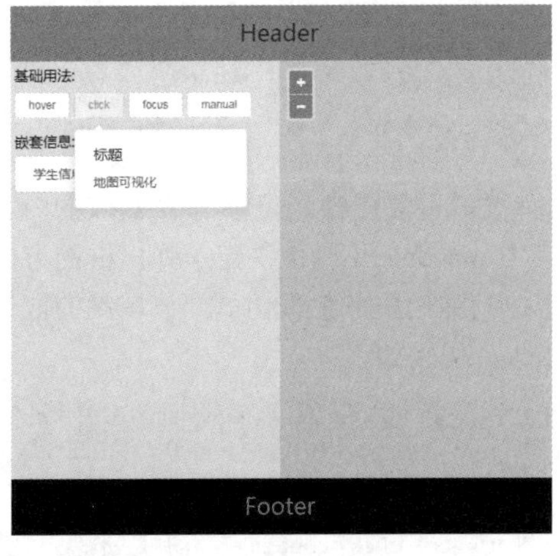

图 7-8　弹出框的样式（一）

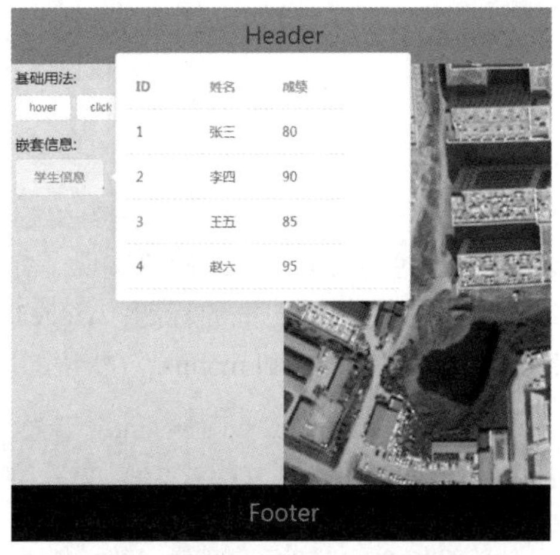

图 7-9　弹出框的样式（二）

```
var vm = new Vue({
    el: "#app",
    data: {
        visible: false,
        tableData: [{
            ID: 1,
            name: "张三",
            grade: 80
        }, {
            ID: 2,
            name: "李四",
            grade: 90
        }, {
            ID: 3,
            name: "王五",
            grade: 85
        }, {
            ID: 4,
            name: "赵六",
            grade: 95
        }]
    },
    methods: {
        //单击基础用法中“click”按钮时的响应事件
        handleClick: function () {
            //获取当前图层的可见性
            var visible = imageLayer.getVisible();
            //若地图当前可见，则变为不可见；若当前不可见，则变为可见
            imageLayer.setVisible(!visible);
        },
        //单击基础用法中“manual”按钮时的响应事件
        handleManualDblclick: function () {
            this.visible = !this.visible;
        }
    }
})
```

（1）基础用法。弹出框与文字提示类似，因此有很多重复属性，可参考 7.2 节的内容。Element 通过封装 el-popover 标签实现了弹出框，其中 placement 属性决定了弹出框的方向位置；title 属性决定了弹出框的标题；trigger 属性用于控制弹出框的触发方式，该属性的取值包括 hover、click、focus 和 manual，默认的取值是 click。代码如下：

```
<el-popover placement="top-start" title="标题" width="100" trigger="hover" content="这是一段内容">
    <el-button slot="reference" size="mini">hover</el-button>
</el-popover>
<el-popover placement="top-start" title="标题" width="100" trigger="click" content="地图可视化">
    <el-button slot="reference" size="mini" @click="handleClick">click</el-button>
```

```
    </el-popover>
    <el-popover placement="top-start" title="标题" width="100" trigger="focus" content="这是一段内容">
        <el-button slot="reference" size="mini">focus</el-button>
    </el-popover>
    <el-popover placement="top-start" title="标题" width="100" trigger="manual" content="这是一段内容"
        v-model="visible">
        <el-button slot="reference" size="mini" @dblclick.native="handleManualDblclick">manual
        </el-button>
    </el-popover>
```

（2）嵌套信息。el-popover 标签中的内容可以是任意的，甚至可以是表格（只需要在 el-popover 标签中嵌入表格即可）。代码如下：

```
    <el-popover placement="right" width="300" trigger="click">
        <el-table :data="tableData">
            <el-table-column prop="ID" label="ID" width="80">
            </el-table-column>
            <el-table-column prop="name" label="姓名" width="80">
            </el-table-column>
            <el-table-column prop="grade" label="成绩" width="80">
            </el-table-column>
        </el-table>
        <el-button slot="reference">学生信息</el-button>
    </el-popover>
```

7.4 卡片（Card）

Element 为卡片提供了不同的样式与用法，本节将从基础用法、带图片、卡片阴影等方面讲解卡片的样式与用法。本节首先结合地图展示卡片的样式，如图 7-10 所示；然后对卡片的各个样式与用法进行讲解。

图 7-10 卡片的样式

本示例首先创建 Vue 实例，methods 中的 handleClick1 是单击基础用法中“地图放大”按钮时的响应事件，handleClick2 是单击基础用法中“地图缩小”按钮时的响应事件；然后单击“地图缩小”按钮或单击“地图放大”按钮，实现地图的缩放。代码如下：

```
var vm = new Vue({
    el: "#app",
    data: {
    },
    methods: {
        //单击基础用法中“地图放大”按钮时的响应事件
        handleClick1: function () {
            var zoom = map.getView().getZoom();
            map.getView().setZoom(zoom + 1);
        },
        //单击基础用法中“地图缩小”按钮时的响应事件
        handleClick2: function () {
            var zoom = map.getView().getZoom();
            map.getView().setZoom(zoom - 1);
        }
    }
})
```

（1）基础用法。卡片的作用是将信息聚合在一个区域内显示。Element 通过对 el-card 标签的封装实现了卡片，卡片包括 header 和 body 部分，header 需要具名的 slot 来分发。代码如下：

```
<el-card>
    <div slot="header">
        <span>校园信息</span>
        <el-button style="float: right;" type="primary" size="mini" @click="handleClick1">地图放大
        </el-button>
        <el-button style="float: right;" type="primary" size="mini" @click="handleClick1">地图缩小
</el-button>
    </div>
    <div>教学楼</div>
    <div>图书馆</div>
    <div>食堂</div>
</el-card>
```

（2）带图片。如果需要在卡片中显示图片，则只需要在 el-card 标签中嵌入 img 标签即可。代码如下：

```
<el-card>
    <img src="./avatar.jpg" alt="">
    <span>你好，我是程序狗</span>
</el-card>
```

（3）卡片阴影。在 el-card 标签中设置 shadow 属性可以控制卡片的阴影显示方式，该属性的取值包括 always、hover、never，默认的取值为 always。代码如下：

```
<el-card shadow="always">A 卡片</el-card>
<el-card shadow="hover">B 卡片</el-card>
<el-card shadow="never">C 卡片</el-card>
```

7.5 走马灯（Carousel）

Element 为走马灯提供了不同的样式与用法，本节将从基础用法、指示器、切换箭头、卡片化、方向等方面讲解走马灯的样式与用法。本节首先结合地图展示走马灯的样式，如图 7-11 所示；然后对走马灯的各个样式与用法进行讲解。

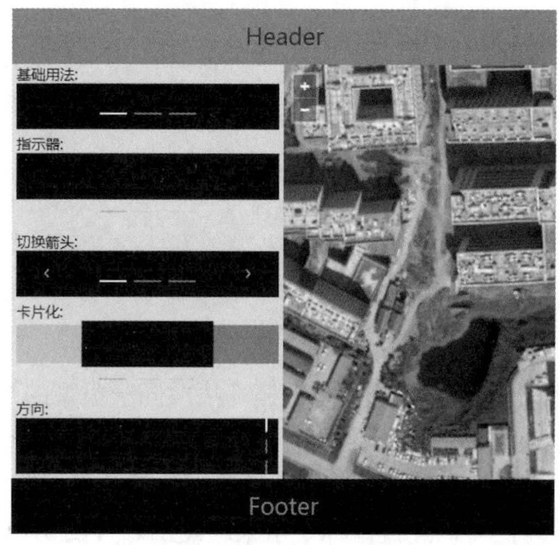

图 7-11　走马灯的样式

本示例创建了 Vue 实例，methods 中的 handleChange 是基础用法中走马灯切换的响应事件，该事件处理方法的参数为幻灯片的索引，此时地图的透明度随着幻灯片的切换而变化，如图 7-12 所示。代码如下：

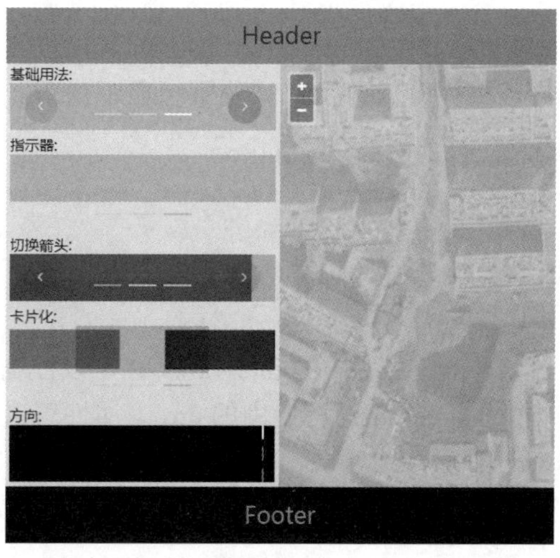

图 7-12　地图的透明度随着幻灯片的切换而变化

```
var vm = new Vue({
    el: "#app",
    data: {
    },
    methods: {
        //基础用法中走马灯切换的响应事件
        handleChange:function(index){
            switch(index){
                case 0:
                    imageLayer.setOpacity(1);
                break;
                case 1:
                    imageLayer.setOpacity(0.6);
                break;
                case 2:
                    imageLayer.setOpacity(0.2);
                break;
            }
        }
    }
})
```

（1）基础用法。走马灯是指在有限空间内，循环播放同一类型的图片、文字等内容，在前端开发中多应用于轮播图。在 el-carousel 标签中嵌入 el-carousel-item 标签可实现走马灯的效果。幻灯片的内容是任意的，在默认情况下，在鼠标光标悬浮在底部的指示器上时会触发幻灯片的切换。将 trigger 属性设置为 click 时，可以通过鼠标单击来触发幻灯片的切换。代码如下：

```
<el-carousel height="50px" @change="handleChange">
    <el-carousel-item class="first"></el-carousel-item>
    <el-carousel-item class="second"></el-carousel-item>
    <el-carousel-item class="third"></el-carousel-item>
</el-carousel>
```

（2）指示器。将 el-carousel 标签中的 indicator-position 属性设置为 outside，可以使指示器的位置显示在外部；如果将该属性设置为 none，则不会显示指示器。代码如下：

```
<el-carousel height="50px" indicator-position="outside">
    <el-carousel-item class="first"></el-carousel-item>
    <el-carousel-item class="second"></el-carousel-item>
    <el-carousel-item class="third"></el-carousel-item>
</el-carousel>
```

（3）切换箭头。通过 el-carousel 标签中的 interval 属性可以控制幻灯片的切换时间，切换时间的单位为毫秒（ms）；通过 el-carousel 标签中的 arrow 属性可以控制走马灯的箭头显示方式，该属性的取值包括 always、hover、never，默认的取值是 hover，always 表示箭头一直显示，hover 表示鼠标光标悬浮在幻灯片的某箭头上时显示该箭头，never 表示箭头永远也不显示。代码如下：

```
<el-carousel height="50px" :interval="1000" arrow="always">
    <el-carousel-item class="first"></el-carousel-item>
    <el-carousel-item class="second"></el-carousel-item>
    <el-carousel-item class="third"></el-carousel-item>
</el-carousel>
```

（4）卡片化。当页面宽度方向的空间有空余，而高度方向的空间匮乏时，可使用卡片，此时只需要将 el-carousel 标签中的 type 属性设置为 card 即可。代码如下：

```
<el-carousel height="50px" type="card">
    <el-carousel-item class="first"></el-carousel-item>
    <el-carousel-item class="second"></el-carousel-item>
    <el-carousel-item class="third"></el-carousel-item>
</el-carousel>
```

（5）方向。将 el-carousel 标签中的 direction 属性设置为 vertical，可以使走马灯在垂直方向上显示。代码如下：

```
<el-carousel height="60px" direction="vertical">
    <el-carousel-item class="first"></el-carousel-item>
    <el-carousel-item class="second"></el-carousel-item>
    <el-carousel-item class="third"></el-carousel-item>
</el-carousel>
```

7.6　折叠面板（Collapse）

Element 为折叠面板提供了不同的样式与用法，本节将从基础用法、手风琴效果、自定义面板标题等方面讲解折叠面板的样式与用法。本节首先结合地图展示折叠面板的样式，如图 7-13 所示；然后对折叠面板的各个样式与用法进行讲解。

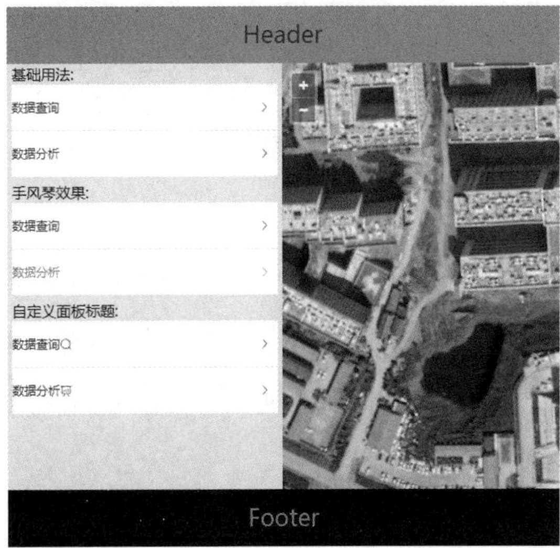

图 7-13　折叠面板的样式

本示例首先创建 Vue 实例，data 中的 activeNames1、activeNames2、activeNames3 分别是各个 el-collapse 标签中 v-model 属性绑定的变量，methods 中的 handleChange 是单击手风琴效果中折叠面板时的响应事件，该事件处理方法的参数是 el-collapse-item 标签中 name 属性所对应的值；然后单击手风琴效果的数据查询和数据分析，地图上会弹出相应的警告，如图 7-14 所示。代码如下：

图 7-14　地图上弹出的警告

```
var vm = new Vue({
    el: "#app",
    data: {
        activeNames1: [],
        activeNames2: [],
        activeNames3: []
    },
    methods: {
        //单击手风琴效果中折叠面板时的响应事件
        handleChange: function (name) {
            if(name==="1"){
             this.$message("数据查询")
            }
            else if(name==="2"){
             this.$message("数据分析")
            }
        }
    }
})
```

（1）基础用法。Element 通过在 el-collapse 标签中嵌入 el-collapse-item 标签实现了折叠面

板，在默认情况下，可同时打开多个面板，面板之间互不影响。el-collapse 标签中 v-model 属性绑定的变量用于监听 el-collapse-item 标签中 name 属性的变化。代码如下：

```
<el-collapse v-model="activeNames1">
    <el-collapse-item title="数据查询" name="1">
        <div>
            <el-button type="text">空间查询</el-button>
        </div>
        <div>
            <el-button type="text">属性查询</el-button>
        </div>
    </el-collapse-item>
    <el-collapse-item title="数据分析" name="2">
        <div>
            <el-button type="text">密度分析</el-button>
        </div>
        <div>
            <el-button type="text">热力图展示</el-button>
        </div>
    </el-collapse-item>
</el-collapse>
```

（2）手风琴效果。手风琴效果是指各个面板之间是互斥的，每次只能打开一个面板，只需要在 el-collapse 标签中添加 accordion 属性即可实现手风琴效果。代码如下：

```
<el-collapse v-model="activeNames2" accordion @change="handleChange">
    <el-collapse-item title="数据查询" name="1">
        <div>
            <el-button type="text">空间查询</el-button>
        </div>
        <div>
            <el-button type="text">属性查询</el-button>
        </div>

    </el-collapse-item>
    <el-collapse-item title="数据分析" name="2">
        <div>
            <el-button type="text">密度分析</el-button>
        </div>
        <div>
            <el-button type="text">热力图展示</el-button>
        </div>
    </el-collapse-item>
</el-collapse>
```

（3）自定义面板标题。通常，Element 是通过 el-collapse-item 标签中的 title 属性来设置面板标题的，若需要设置更加复杂的标题，如在标题后添加图标，那么就需要使用自定义面板标题，通过 template 标签和 slot 属性可以实现自定义面板标题。代码如下：

```
<el-collapse v-model="activeNames3" accordion>
    <el-collapse-item name="1">
        <template slot="title">
            数据查询
            <i class="el-icon-search"></i>
        </template>
        <div>
            <el-button type="text">空间查询</el-button>
        </div>
        <div>
            <el-button type="text">属性查询</el-button>
        </div>
    </el-collapse-item>
    <el-collapse-item name="2">
        <template slot="title">
            数据分析
            <i class="el-icon-data-analysis"></i>
        </template>
        <div>
            <el-button type="text">密度分析</el-button>
        </div>
        <div>
            <el-button type="text">热力图展示</el-button>
        </div>
    </el-collapse-item>
</el-collapse>
```

7.7 时间线（Timeline）

Element 为时间线提供了不同的样式与用法，本节将从基础用法、定义时间戳两个方面讲解时间线的样式与用法。本节首先结合地图展示时间线的样式，如图 7-15 所示；然后对时间线的各个样式与用法进行讲解。

本示例首先创建 Vue 实例，data 中的 reverse 与基础用法中的 el-radio-group 标签中的 v-model 属性、el-timeline 标签中的 reverse 属性相关联，用于控制时间线的排序，methods 中的 handleChange 是单选框变化时的响应事件；然后单击基础用法中的“倒序”和“正序”单选框，会弹出第一个时间戳，如图 7-16 所示。代码如下：

```
var vm = new Vue({
    el: "#app",
    data: {
        reverse: false
    },
    methods: {
```

```
            //单选框变化时的响应事件
            handleChange: function () {
                this.$message(this.$refs.date.$children[0].timestamp)
            }
        }
    })
```

图 7-15　时间线的样式

图 7-16　弹出的第一个时间戳

（1）基础用法。Element 通过在 el-ctimeline 标签中嵌入 el-timeline-item 标签提供了时间线。el-ctimeline 标签中的 reverse 属性用于控制时间线的排序，ref 属性用于获取时间线标签；el-timeline-item 标签中的 timestamp 是时间戳，color 是节点的颜色，icon 是图标，size 用于控制尺寸。代码如下：

```
<el-radio-group v-model="reverse" @change="handleChange">
    <el-radio :label="true">倒序</el-radio>
    <el-radio :label="false">正序</el-radio>
</el-radio-group>
<el-timeline :reverse="reverse" ref="date">
    <el-timeline-item timestamp='2018-09-15' color="blue" icon="el-icon-school" size="large">
        开学日期</el-timeline-item>
    <el-timeline-item timestamp='2019-01-15'>放假日期</el-timeline-item>
</el-timeline>
```

（2）定义时间戳。在 el-timeline-item 标签中嵌入其他标签可以定义时间戳。代码如下：

```
<el-timeline>
    <el-timeline-item timestamp="2018/9/15" placement="top">
        <el-card>
            <p>开学时间到了!</p>
        </el-card>
    </el-timeline-item>
```

```
    <el-timeline-item timestamp="2019/1/15" placement="top">
        <el-card>
            <p>放假时间到了!</p>
        </el-card>
    </el-timeline-item>
</el-timeline>
```

7.8 分割线（Divider）

Element为分割线提供了不同的样式与用法，本节将从基础用法、设置文案、垂直分割等方面讲解分割线的样式与用法。本节首先结合地图展示分割线的样式，如图7-17所示；然后对分割线的各个样式与用法进行讲解。

图7-17　分割线的样式

本示例创建了Vue实例，由于本示例比较简单，data和methods都为空。代码如下：

```
var vm = new Vue({
    el: "#app",
    data: {
    },
    methods: {
    }
})
```

（1）基础用法。分割线是用于区分内容的线条，Element通过el-divider标签提供了分割线。代码如下：

```
<div>
    <span>这是第一行文字</span>
    <el-divider></el-divider>
```

```
    <span>这是第二行文字</span>
</div>
```

（2）设置文案。用户可以在分割线上自定义文案内容，通过 el-divider 标签中的 content-position 属性可以控制文案的位置。代码如下：

```
<el-divider content-position="left">这是一段文字</el-divider>
<el-divider content-position="center">这是一段文字</el-divider>
<el-divider content-position="right">这是一段文字</el-divider>
```

（3）垂直分割。将 el-divider 标签中的 direction 属性设置为 vertical，可以实现垂直分割。代码如下：

```
<span>First</span>
<el-divider direction="vertical"></el-divider>
<span>Second</span>
<el-divider direction="vertical"></el-divider>
<span>Third</span>
```

7.9　日历（Calendar）

Element 为日历提供了不同的样式与用法，本节将从基础用法、自定义内容、自定义范围等方面讲解日历的样式和用法。

由于日历所占的物理空间较大，因此本节分别用一个页面布局对每个样式进行讲解。

（1）基础用法。Element 通过 el-calendar 标签提供了日历，el-calendar 标签中的 v-model 属性所对应的变量为当前日期。

本示例首先创建 Vue 实例，data 中的 value 为当前日期；然后结合地图展示基础用法的样式，如图 7-18 所示。代码如下：

图 7-18　基础用法的样式

```
var vm = new Vue({
    el: "#app",
    data: {
        value:new Date
    },
    methods: {
    }
})
```

基础用法样式对应的标签代码如下：

```
<el-calendar v-model="value"></el-calendar>
```

（2）自定义内容。在 el-calendar 标签中设置 template 标签可以自定义日历的内容。需要注意的是，在自定义内容时，首先需要将 slot 属性设置为 dateCell，然后通过 scoped-slot 属性获取 date，date 就是当前单元格的日期。

本示例首先创建 Vue 实例，data 中的 value 是当前日期；然后结合地图展示自定义内容的样式，如图 7-19 所示。在本示例中，当单击某个日期，该日期下方会显示一个勾子。代码如下：

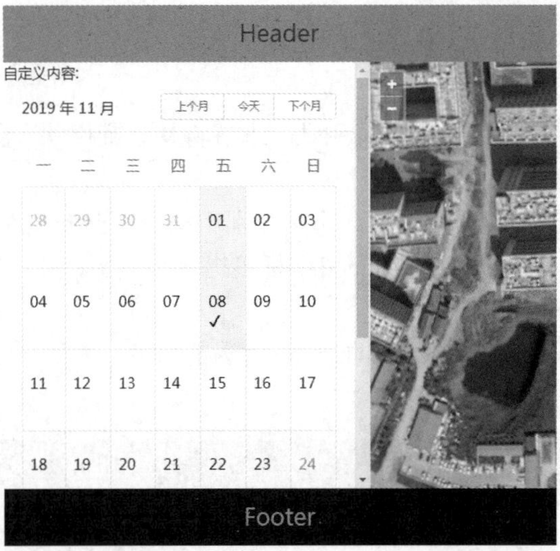

图 7-19　自定义内容的样式

```
var vm = new Vue({
    el: "#app",
    data: {
        value:new Date
    },
    methods: {
    }
})
```

自定义内容样式对应的标签代码如下：

```
<el-calendar>
```

```
        <template slot="dateCell" slot-scope="{date, data}">
            <p>
                {{ data.day.split('-').slice(2).join('') }}
                {{ data.isSelected ? '✔️' : ''}}
            </p>
        </template>
    </el-calendar>
```

（3）自定义范围。在 el-calendar 标签中设置 range 属性可以自定义日历的选择范围。本示例首先创建 Vue 实例，data 中的 value 为当前日期；然后结合地图展示自定义范围的样式，如图 7-20 所示。代码如下：

图 7-20　自定义范围的样式

```
    var vm = new Vue({
        el: "#app",
        data: {
            value:new Date
        },
        methods: {
        }
    })
```

自定义范围样式对应的标签代码如下：

```
    <el-calendar v-model="value" :range="['2019-11-04', '2019-11-24']"></el-calendar>
```

7.10　图片（Image）

Element 为图片提供了不同的样式与用法，本节将从基础用法、加载失败、懒加载等方面讲解图片的样式与用法。本节首先结合地图展示图片的样式，如图 7-21 所示；然后对图片的各个样式与用法进行讲解。

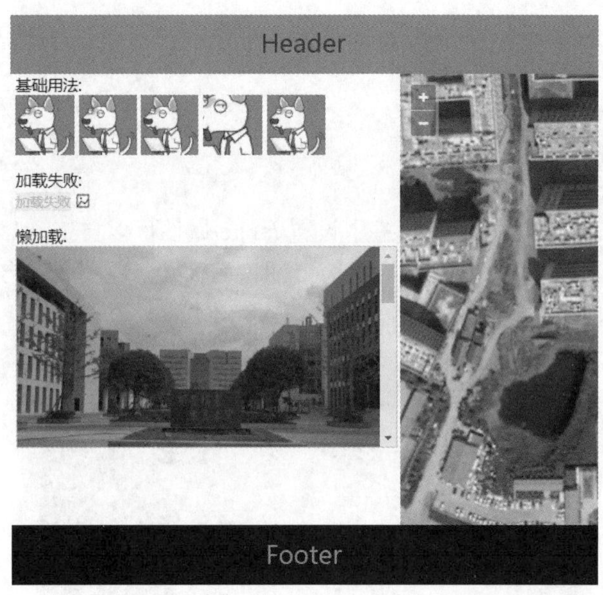

图 7-21　图片的样式

本示例创建了 Vue 实例，由于本示例比较简单，data 和 methods 都为空。代码如下：

```
var vm = new Vue({
    el: "#app",
    data: {
    },
    methods: {
    }
})
```

（1）基础用法。Element 通过 el-image 标签提供了图片。el-image 标签中的 src 属性是图片路径。el-image 标签中的 fit 属性用于决定图片占用容器框的方式，该属性的取值包括 fill、contain、cover、none 和 scale-down，默认的取值是 cover，其中 fill 表示将图片拉伸充满整个容器框；contain 表示保持图片原有的尺寸比例，长度和高度中较短的那条边和容器大小一致，较长的那条边等比例缩放，可能会留有空白；cover 表示保持图片原有的尺寸比例，宽度和高度中较长的那条边和容器大小一致，较短的那条边等比例缩放，可能留有空白；none 表示保持图片原有的尺寸比例；scale-down 表示保持图片原有的尺寸比例，如果容器尺寸大于图片尺寸，则保持图片的原有尺寸，不会放大失真，如果容器尺寸小于图片尺寸，则用法跟 contain 一样。代码如下：

```
<el-image style="width: 60px; height: 60px" src="./avatar.jpg" fit="fill"></el-image>
<el-image style="width: 60px; height: 60px" src="./avatar.jpg" fit="contain"></el-image>
<el-image style="width: 60px; height: 60px" src="./avatar.jpg" fit="cover"></el-image>
<el-image style="width: 60px; height: 60px" src="./avatar.jpg" fit="none"></el-image>
<el-image style="width: 60px; height: 60px" src="./avatar.jpg" fit="scale-down"></el-image>
```

（2）加载失败。在图片加载失败时通常需要显示加载失败，将 el-image 标签中的 slot 属性设置为 error，可以自定义图片加载失败时显示的内容。代码如下：

```
<el-image></el-image>
<el-image>
    <div slot="error" class="image-slot">
        <i class="el-icon-picture-outline"></i>
    </div>
</el-image>
```

（3）懒加载。在 el-image 标签中添加 lazy 属性可以实现图片的懒加载。代码如下：

```
<div style="width: 390px;height: 200px;border: 1px solid #ccc; overflow: auto;">
    <el-image  src="./1.jpg" lazy></el-image>
    <el-image  src="./2.jpg" lazy></el-image>
    <el-image  src="./3.jpg" lazy></el-image>
</div>
```

7.11　无限滚动（InfiniteScroll）

Element 提供了无限滚动的样式与用法，本节只对无限滚动的基础用法进行讲解。无限滚动的样式如图 7-22 所示。

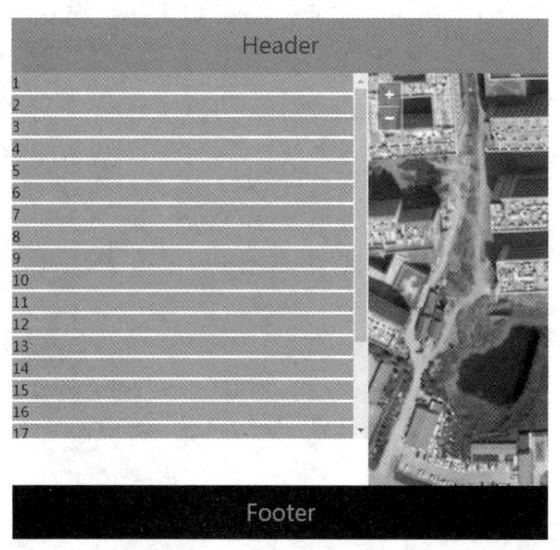

图 7-22　无限滚动的样式

本示例创建了 Vue 实例，data 中的 count 是列表渲染的总数，methods 中的 load 是滚动条滚动到底部时的响应事件。代码如下：

```
var vm = new Vue({
    el: "#app",
    data: {
```

```
            count: 0
        },
        methods: {
            load:function() {
                this.count +=1
            }
        },
    })
```

在列表中添加 v-infinite-scroll，并赋值相应的加载方法，当滚动到列表底部时可自动执行加载方法。代码如下：

```
<ul class="list" v-infinite-scroll="load" style="overflow:auto">
    <li v-for="i in count" class="list-item">{{ i }}</li>
</ul>
```

7.12 抽屉（Drawer）

Element 为抽屉提供了不同的样式与用法，本节将从基础用法、自定义内容、多层嵌套等方面讲解抽屉的样式与用法。本节首先结合地图展示与抽屉相关的按钮布局，如图 7-23 所示；然后对抽屉的各个样式与用法进行讲解。

图 7-23　与抽屉相关的按钮布局

本示例首先创建 Vue 实例，data 中的 drawer1 至 drawer3_2 与各个抽屉标签中的 visible.sync 属性相关联，direction1 至 direction3_2 与各个抽屉标签中的 direction 属性相关联，tableData 与 el-table 标签中的 data 属性相关联，methods 中的 handleClick1 是单击基础用法中“打开”按钮时的响应事件，handleClick2 是单击自定义内容中“打开表格”按钮时的响应事件，

handleClick3_1 是单击多层嵌套中“打开嵌套抽屉”按钮时的响应事件，handleClick3_2 是单击父抽屉中“打开子抽屉”按钮时的响应事件。代码如下：

```
var vm = new Vue({
    el: "#app",
    data: {
        drawer1: false,
        direction1: 'rtl',
        drawer2: false,
        direction2: 'rtl',
        drawer3_1: false,
        direction3_1: 'rtl',
        drawer3_2: false,
        direction3_2: 'rtl',
        tableData: [{
            ID: 1,
            name: "张三",
            grade: 80
        }, {
            ID: 2,
            name: "李四",
            grade: 90
        }, {
            ID: 3,
            name: "王五",
            grade: 85
        }, {
            ID: 4,
            name: "赵六",
            grade: 95
        }]
    },
    methods: {
        //单击基础用法中“打开”按钮时的响应事件
        handleClick1: function () {
            this.drawer1 = true
        },
        //单击自定义内容中“打开表格”按钮时的响应事件
        handleClick2: function () {
            this.drawer2 = true
        },
        //单击多层嵌套中“打开嵌套抽屉”按钮时的响应事件
        handleClick3_1: function () {
            this.drawer3_1 = true
        },
        //单击父抽屉中“打开子抽屉”按钮时的响应事件
        handleClick3_2: function () {
```

```
                this.drawer3_2 = true
            }
        },
    })
```

（1）基础用法。基础用法的样式如图 7-24 所示。在某些应用场景中，对话框无法满足实际的需求，如需要很长的表单或者需要临时展示一些文档的应用场景，这时抽屉可以很好地满足实际的需求。Element 是通过 el-drawer 标签提供抽屉的，el-drawer 标签的 title 属性是抽屉的标题；visible.sync 属性用来控制抽屉的可视性；direction 属性用来控制开抽屉的方向，该属性的取值包括 rtl、ltr、ttb、btt，分别表示从右往左开、从左往右开、从上往下开、从下往上开，默认的取值是 rtl。代码如下：

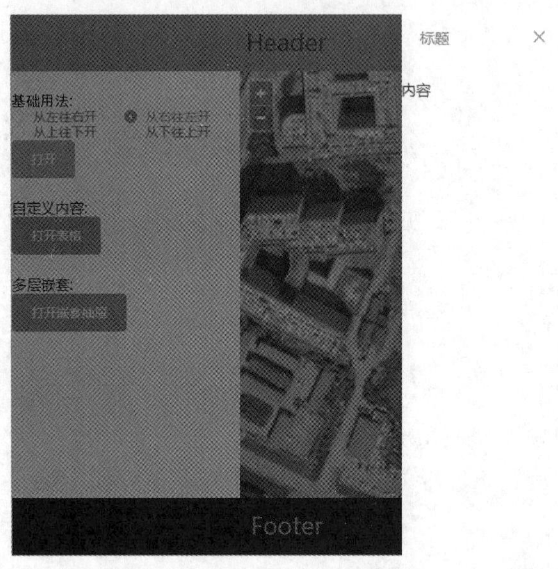

图 7-24　基础用法的样式

```
<el-radio-group v-model="direction1">
    <el-radio label="ltr">从左往右开</el-radio>
    <el-radio label="rtl">从右往左开</el-radio>
    <el-radio label="ttb">从上往下开</el-radio>
    <el-radio label="btt">从下往上开</el-radio>
</el-radio-group>
<el-button @click="handleClick1" type="primary">
    打开
</el-button>
<el-drawer title="标题" :visible.sync="drawer1" :direction="direction1">
    <span>内容</span>
</el-drawer>
```

（2）自定义内容。自定义内容的样式如图 7-25 所示。在 el-drawer 标签中嵌入 el-table 标签可以在抽屉中展示表格。单击自定义内容中的“打开表格”按钮，可弹出带有表格的抽屉。代码如下：

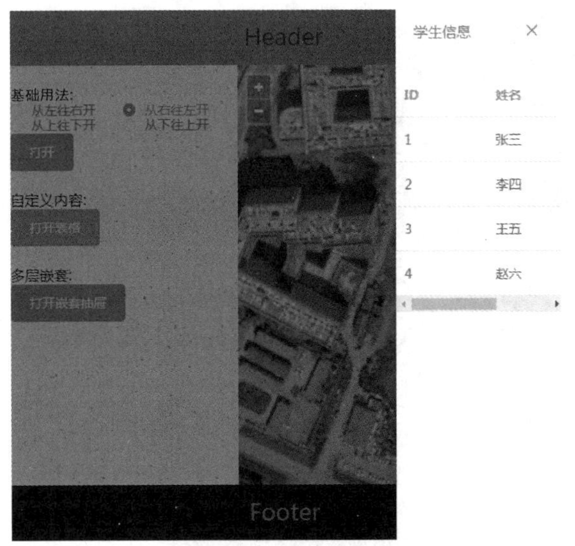

图 7-25　自定义内容的样式

```
<el-button @click="handleClick2" type="primary">
    打开表格
</el-button>
<el-drawer title="学生信息" :visible.sync="drawer2" :direction="direction2">
    <el-table :data="tableData">
        <el-table-column prop="ID" label="ID" width="100">
        </el-table-column>
        <el-table-column prop="name" label="姓名" width="100">
        </el-table-column>
        <el-table-column prop="grade" label="成绩" width="100">
        </el-table-column>
    </el-table>
</el-drawer>
```

（3）多层嵌套。多层嵌套的样式如图 7-26 所示。用户不仅可以在抽屉内嵌套表格，还可以在抽屉内嵌套抽屉。需要注意的是，在嵌套抽屉时，需要在子抽屉中将 append-to-body 属性设置为 true。单击多层嵌套中的“打开嵌套抽屉”按钮，可弹出父抽屉；单击父抽屉中的“打开子抽屉”按钮，可弹出子抽屉。代码如下：

```
<div>多层嵌套:</div>
<el-button @click="handleClick3_1" type="primary">
    打开嵌套抽屉
</el-button>
<el-drawer title="父抽屉" :visible.sync="drawer3_1" :direction="direction3_1" size="50%">
    <el-button @click="handleClick3_2">打开子抽屉</el-button>
    <el-drawer title="子抽屉" :append-to-body="true"
    :visible.sync="drawer3_2" :direction="direction3_2">
    </el-drawer>
</el-drawer>
```

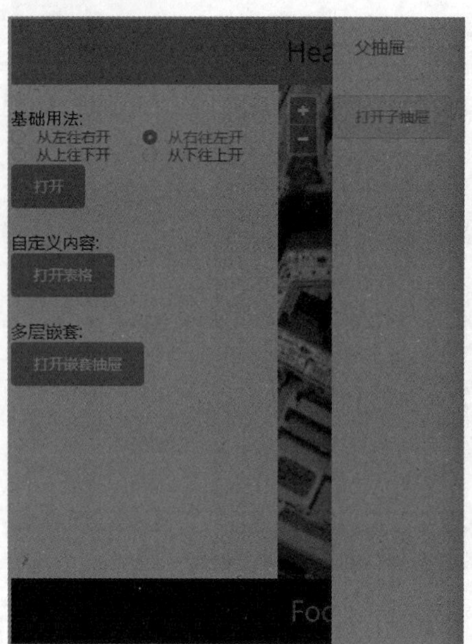

图 7-26　多层嵌套的样式

7.13　思考与练习题

（1）参考 7.1 节，编写一个对话框，分析对话框与通知组件的区别和联系。

（2）参考 7.2 节和 7.3 节，编写文字提示与弹出框，分析它们的区别和联系。

（3）参考 4.6 节，编写折叠面板，分析它与导航菜单的区别和联系。

第 8 章 Element+Vue+OpenLayers 项目实战

本章将将带领读者进行项目实战，开发一个简单的智慧校园系统。为了能够适应主流的 GIS 平台，因此智慧校园系统不指定 GIS 平台和数据库，该系统涉及的数据可参考配套资料中的代码。

8.1 智慧校园系统的需求分析

智慧校园系统除了需要具有地图的一些基本功能，还需要具有可以查询主要地物的功能，即单击某一按钮，会高亮显示该按钮对应的地物并弹出该地物的属性框。由于智慧校园系统的地图是一张图片，因此可以将各个地物的坐标数据直接写在代码中。如果读者使用的是其他 GIS 平台，则可以直接调用相应的服务。

8.2 智慧校园系统的设计

8.2.1 开发环境

智慧校园系统是基于 WebGIS 开发的，不仅使用了最基本的“前端三剑客”——HTML5、CSS3、JavaScript，也使用了 Vue 框架与 Element 组件，还结合 OpenLayers 实现了地图的一些基本功能和查询功能。智慧校园系统使用的编辑器为 Visual Studio Code，开发环境的操作系统是 Windows 7，开发环境的配置请参考 1.1 节的内容，本章使用命令行的方式来创建 Vue 项目。

8.2.2 数据结构设计

智慧校园系统涉及的数据主要是各个地物的空间数据及其属性数据。空间数据主要分为点、线、面的坐标。本节主要介绍各个地物的属性数据，在设置地物的属性数据时，本章将前面学到的 Element 封装的时间选择器、日期范围选择器、开关、计数器等组件和表格组件结合在一起，可以巩固 Element 的学习。

（1）路灯。路灯的属性数据表结构如表 8-1 所示。

表 8-1　路灯的属性数据表结构

字　段　名	数 据 类 型	说　　明
ID	整型	地物的 ID
编号	字符串	地物的编号
名称	字符串	地物的名称
管理单位	字符串	管理地物的部门
负责人	字符串	管理地物的负责人
负责人手机	整型	负责人的手机号码
开启时间	对象	路灯开启的时间，用 Element 组件表示
关闭时间	对象	路灯关闭的时间，用 Element 组件表示

（2）摄像头。摄像头的属性数据表结构如表 8-2 所示。

表 8-2　摄像头的属性数据表结构

字　段　名	数 据 类 型	说　　明
ID	整型	地物的 ID
编号	字符串	地物的编号
名称	字符串	地物的名称
管理单位	字符串	管理地物的部门
负责人	字符串	管理地物的负责人
负责人手机	整型	负责人的手机号码
运行状态	对象	摄像头的运行状态，用 Element 组件表示

（3）指路牌。指路牌的属性数据表结构如表 8-3 所示。

表 8-3　指路牌的属性数据表结构

字　段　名	数 据 类 型	说　　明
ID	整型	地物的 ID
编号	字符串	地物的编号
名称	字符串	地物的名称
管理单位	字符串	管理地物的部门
负责人	字符串	管理地物的负责人
负责人手机	整型	负责人的手机号码

（4）道路。道路的属性数据表结构如表 8-4 所示。

表 8-4 道路的属性数据表结构

字 段 名	数 据 类 型	说 明
ID	整型	地物的 ID
编号	字符串	地物的编号
名称	字符串	地物的名称
管理单位	字符串	管理地物的部门
负责人	字符串	管理地物的负责人
负责人手机	整型	负责人的手机号码
长度	整型	道路的长度
是否允许临时停车	布尔型	用 Element 组件表示

（5）足球场。足球场的属性数据表结构如表 8-5 所示。

表 8-5 足球场的属性数据表结构

字 段 名	数 据 类 型	说 明
ID	整型	地物的 ID
编号	字符串	地物的编号
名称	字符串	地物的名称
管理单位	字符串	管理地物的部门
负责人	字符串	管理地物的负责人
负责人手机	整型	负责人的手机号码
开放时间段	对象	开放的时间范围，用 Element 组件表示
今日是否开放	布尔型	用 Element 组件表示

（6）篮球场。篮球场的属性数据表结构如表 8-6 所示。

表 8-6 篮球场的属性数据表结构

字 段 名	数 据 类 型	说 明
ID	整型	地物的 ID
编号	字符串	地物的编号
名称	字符串	地物的名称
管理单位	字符串	管理地物的部门
负责人	字符串	管理地物的负责人
负责人手机	整型	负责人的手机号码
场地数量	整型	篮球场的场地数量
开放时间段	对象	开放的时间范围，用 Element 组件表示
今日是否开放	布尔型	用 Element 组件表示

（7）体育馆。体育馆的属性数据表结构如表 8-7 所示。

表 8-7 体育馆的属性数据表结构

字 段 名	数 据 类 型	说 明
ID	整型	地物的 ID
编号	字符串	地物的编号
名称	字符串	地物的名称
管理单位	字符串	管理地物的部门
负责人	字符串	管理地物的负责人
负责人手机	整型	负责人的手机号码
服务项目	对象	服务的项目类型，用 Element 组件表示
开放时间段	对象	开放的时间范围，用 Element 组件表示
今日是否开放	布尔型	用 Element 组件表示

（8）湖泊。湖泊的属性数据表结构如表 8-8 所示。

表 8-8 湖泊的属性数据表结构

字 段 名	数 据 类 型	说 明
ID	整型	地物的 ID
编号	字符串	地物的编号
名称	字符串	地物的名称
管理单位	字符串	管理地物的部门
负责人	字符串	管理地物的负责人
负责人手机	整型	负责人的手机号码
面积	字符串	湖泊的面积
水深	字符串型	湖泊的深度

（9）公教楼。公教楼的属性数据表结构如表 8-9 所示。

表 8-9 公教楼的属性数据表结构

字 段 名	数 据 类 型	说 明
ID	整型	地物的 ID
编号	字符串	地物的编号
名称	字符串	地物的名称
管理单位	字符串	管理地物的部门
负责人	字符串	管理地物的负责人
负责人手机	整型	负责人的手机号码
楼层数	整型	公教楼的楼层数，用 Element 组件表示
教室数量	整型	公教楼的教室数，用 Element 组件表示

（10）图书馆。图书馆的属性数据表结构如表 8-10 所示。

表 8-10　图书馆的属性数据表结构

字　段　名	数 据 类 型	说　　明
ID	整型	地物的 ID
编号	字符串	地物的编号
名称	字符串	地物的名称
管理单位	字符串	管理地物的部门
负责人	字符串	管理地物的负责人
负责人手机	整型	负责人的手机号码
楼层数	整型	图书馆的楼层数，用 Element 组件表示
阅读室数量	整型	阅读室数量，用 Element 组件表示
服务项目	对象	服务的项目类型，用 Element 组件表示
开放时间段	对象	开放的时间范围，用 Element 组件表示

（11）重点实验室。重点实验室的属性数据表结构如表 8-11 所示。

表 8-11　重点实验室的属性数据表结构

字　段　名	数 据 类 型	说　　明
ID	整型	地物的 ID
编号	字符串	地物的编号
名称	字符串	地物的名称
管理单位	字符串	管理地物的部门
负责人	字符串	管理地物的负责人
负责人手机	整型	负责人的手机号码
楼层数	整型	实验室的楼层数，用 Element 组件表示
房间个数	整型	房间个数，用 Element 组件表示
开放时间段	对象	开放的时间范围，用 Element 组件表示

（12）教学服务中心。教学服务中心的属性数据表结构如表 8-12 所示。

表 8-12　教学服务中心的属性数据表结构

字　段　名	数 据 类 型	说　　明
ID	整型	地物的 ID
编号	字符串	地物的编号
名称	字符串	地物的名称
管理单位	字符串	管理地物的部门
负责人	字符串	管理地物的负责人
负责人手机	整型	负责人的手机号码
楼层数	整型	实验室的楼层数，用 Element 组件表示
房间个数	整型	房间个数，用 Element 组件表示
开放时间段	对象	开放的时间范围，用 Element 组件表示

（13）食堂。食堂的属性数据表结构如表 8-13 所示。

表 8-13　食堂的属性数据表结构

字　段　名	数 据 类 型	说　　明
ID	整型	地物的 ID
编号	字符串	地物的编号
名称	字符串	地物的名称
管理单位	字符串	管理地物的部门
负责人	字符串	管理地物的负责人
负责人手机	整型	负责人的手机号码
楼层数	整型	食堂的楼层数，用 Element 组件表示
窗口个数	整型	食堂的窗口个数，用 Element 组件表示
开放时间段	对象	开放的时间范围，用 Element 组件表示

（14）办公楼。办公楼的属性数据表结构如表 8-14 所示。

表 8-14　办公楼的属性数据表结构

字　段　名	数 据 类 型	说　　明
ID	整型	地物的 ID
编号	字符串	地物的编号
名称	字符串	地物的名称
管理单位	字符串	管理地物的部门
负责人	字符串	管理地物的负责人
负责人手机	整型	负责人的手机号码
楼层数	整型	办公楼的楼层数，用 Element 组件表示
办公室个数	整型	办公室的个数，用 Element 组件表示
开放时间段	对象	开放的时间范围，用 Element 组件表示

（15）学生宿舍楼。学生宿舍楼的属性数据表结构如表 8-15 所示。

表 8-15　学生宿舍楼的属性数据表结构

字　段　名	数 据 类 型	说　　明
ID	整型	地物的 ID
编号	字符串	地物的编号
名称	字符串	地物的名称
管理单位	字符串	管理地物的部门
负责人	字符串	管理地物的负责人
负责人手机	整型	负责人的手机号码
楼栋个数	整型	宿舍楼的个数，用 Element 组件表示
楼层数	整型	每个楼栋的楼层数，用 Element 组件表示
开放时间段	对象	开放的时间范围，用 Element 组件表示

8.2.3　系统功能设计

智慧校园系统的功能分为地图基本功能和查询功能。地图基本功能包括地图放大、地图缩小、地图复位和地图平移。查询功能是对各个地物进行查询，智慧校园系统对各个地物进行了分类，主要分为道路设施、运动休闲、教学设施、餐饮服务、办公楼、学生宿舍。智慧校园系统的功能设计如图 8-1 所示。

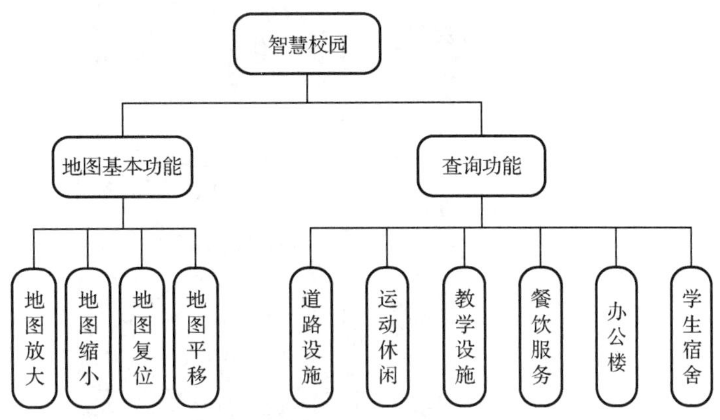

图 8-1　智慧校园系统的功能设计

8.3　智慧校园系统的功能实现

智慧校园系统的主页面如图 8-2 所示，主页面的右上角是实现地图基本功能的 4 个图标，分别是地图放大、地图缩小、地图复位和地图平移。主页面的左侧栏是一个导航菜单，用于实现查询功能。智慧校园系统将地物分为道路设施、运动休闲、教学设施、餐饮服务、办公楼、学生宿舍。

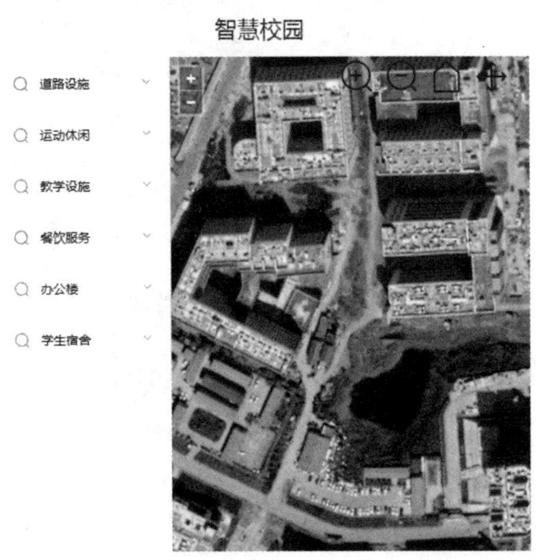

图 8-2　智慧校园系统的主页面

8.3.1 地图基本功能

单击智慧校园系统主页面右上角的图标，可实现相应的地图基本功能。地图放大功能、地图缩小功能、地图复位功能和地图平移功能分别如图 8-3、图 8-4、图 8-5 和图 8-6 所示。

图 8-3 地图放大功能

图 8-4 地图缩小功能

图 8-5 地图复位功能

图 8-6 地图平移功能

8.3.2 道路设施查询

单击导航菜单中的一级菜单“道路设施”，即可显示该菜单下的各个地物，分别为路灯、摄像头、指路牌、道路一、道路二。单击二级菜单“路灯”，对应的地物便可在地图上高亮显示出来，并弹出路灯的属性数据表，如图 8-7 所示。

智慧校园

道路设施
摄像头
指路牌
道路一
道路二
运动休闲
教学设施
餐饮服务
办公楼

ID	负责人手机	开启时间	关闭时间
1	15623211111	18:00:00 ...	18:00:00
2	15623211111	18:00:00 ...	18:00:00
3	15623211111	18:00:00 ...	18:00:00

图 8-7　道路设施查询（一）

单击路灯属性数据表中的任一行（如第 2 行），可弹出一个消息提示框，如图 8-8 所示。

图 8-8　道路设施查询（二）

8.3.3　运动休闲查询

单击导航菜单中的一级菜单“运动休闲”，即可显示该菜单下的各个地物，分别为足球场、篮球场（北）、篮球场（南）、羽毛球场、体育馆、湖泊。单击二级菜单“足球场”，对应的地物便可在地图上高亮显示出来，并弹出足球场的属性数据表，如图 8-9 所示。

图 8-9 运动休息查询

8.3.4 教学设施查询

单击导航菜单中的一级菜单“教学设施”，即可显示该菜单下的各个地物，分别为公教楼 1、公教楼 2、图书馆、重点实验室、教学服务中心。单击二级菜单“图书馆”，对应的地物便可在地图上高亮显示出来，并弹出图书馆的属性数据表，如图 8-10 所示。

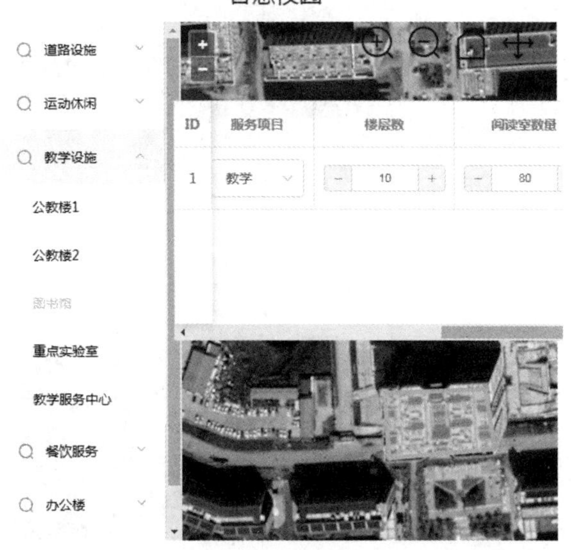

图 8-10 教学设施查询

8.3.5 餐饮服务查询

单击导航菜单中的一级菜单“餐饮服务”，即可显示该菜单下的各个地物，分别为教工食堂、学一食堂、学二食堂。单击二级菜单“学二食堂”，对应的地物便可在地图上高亮显示出来，并弹出学二食堂的属性数据表，如图 8-11 所示。

图 8-11 餐饮服务查询

8.3.6 办公楼查询

单击导航菜单中的一级菜单“办公楼”，即可显示该菜单下的各个地物，分别为材化学院、环境学院、计算机学院、地信学院、经管学院。单击二级菜单“地信学院”，对应的地物便可在地图上高亮显示出来，并弹出地信学院的属性数据表，如图 8-12 所示。

图 8-12 办公楼查询

8.3.7　学生宿舍查询

单击导航菜单中的一级菜单“学生宿舍”，即可显示该菜单下的各个地物，分别为一组团、二组团、三组团、四组团、五组团。单击二级菜单“三组团”，对应的地物便可在地图上高亮显示出来，并弹出三组团的属性数据表，如图 8-13 所示。

图 8-13　学生宿舍查询